# SCIENCE, BELIEF AND SOCIETY

## International Perspectives on Religion, Non-Religion and the Public Understanding of Science

Edited by
Stephen H. Jones,
Tom Kaden and Rebecca Catto

With a Foreword by
Grace Davie

BRISTOL
UNIVERSITY
PRESS

First published in Great Britain in 2019 by

Bristol University Press
University of Bristol
1-9 Old Park Hill
Bristol
BS2 8BB
UK
t: +44 (0)117 954 5940
e: bup-info@bristol.ac.uk
www.bristoluniversitypress.co.uk

North America office:
Policy Press
c/o The University of Chicago Press
1427 East 60th Street
Chicago, IL 60637, USA
t: +1 773 702 7700
f: +1 773-702-9756
e: sales@press.uchicago.edu
www.press.uchicago.edu



British Library Cataloguing in Publication Data
A catalogue record for this book is available from the British Library

Library of Congress Cataloging-in-Publication Data
A catalog record for this book has been requested

ISBN 978-1-5292-0694-4 hardcover
ISBN 978-1-5292-0697-5 ePub
ISBN 978-1-5292-0698-2 Mobi
ISBN 978-1-5292-0696-8 ePdf



Cover design by blu inc
Front cover image: iStock
Printed and bound in Great Britain by CPI Group (UK) Ltd, Croydon, CR0 4YY
Bristol University Press uses environmentally responsible print partners

MIX
Paper from responsible sources
FSC www.fsc.org FSC® C013604

# Contents

# List of Figures and Tables

## Figures

## Tables

# Notes on Contributors

**Jessica Carlisle** is a Research Fellow in the Centre for Visual Cultures at Goldsmiths, University of London. Her focus is law, society and migration in the Middle East and North Africa, especially Syria, Morocco, Egypt and Libya. She is the author of *Shari'a divorce in the Middle East: Contesting gender in the contemporary contract courts* (Palgrave, forthcoming).

**Rebecca Catto** is an Assistant Professor of Sociology at Kent State University. Her research is qualitative in approach and framed by an overarching interest in religious–secular relations: how religious individuals and groups in Europe, North America and beyond engage across boundaries with those of other faiths, no faith and especially in the public realm, and the power dynamics involved. She has published articles on atheism, Christian mission, youth and religion, state–religion relations, and science and religion. She has been a co-investigator on the £1.9 million 'Science and Religion: Exploring the Spectrum' project funded by the Templeton Religion Trust.

**Grace Davie** is Professor Emeritus in the Sociology of Religion at the University of Exeter, UK, and a senior adviser to the Impact of Religion Research Programme at Uppsala University. She is a past-president of the American Association for the Sociology of Religion (2003) and of the Research Committee 22 (Sociology of Religion) of the International Sociological Association (2002–06). In addition to numerous chapters and articles, she is the author of *Religion in Britain since 1945* (Blackwell, 1994), *Religion in modern Europe* (OUP, 2000), *Europe: The exceptional case* (DLT, 2002), *The sociology of religion* (Sage, 2013 [2007]) and *Religion in Britain: A persistent paradox* (Wiley-Blackwell, 2015).

**Fern Elsdon-Baker** is a Professor of Science, Knowledge and Belief in Society at the University of Birmingham, UK. She was principal investigator (PI) on the 'Science and Religion: Exploring the

Spectrum' project between 2014 and 2017. She was previously Head of 'Darwin Now' (2008–10), which ran in 50 countries worldwide. She then went on to work as director of a large-scale multi-region portfolio programme, 'Belief in Dialogue' (2010 –012), which incorporated a number of projects that focused on creating spaces for public, policy and academic dialogue around global issues relating to interfaith dialogue, including exploring the relationship between science, faith and culture.

**Silke Gülker** is a Research Fellow at the Institute for Cultural Studies at the University of Leipzig, Germany. She received her doctoral degree (DPhil) in Political Science from the Free University in Berlin and is an associate in the 'Science, Religion, and Culture Program' at Harvard University, as well as in the Humanities Centre for Advanced Studies' 'Multiple Secularities: Beyond the West, Beyond Modernities' at the University of Leipzig.

**Salman Hameed** is Charles Taylor Chair and Associate Professor of Integrated Science and Humanities in the School of Cognitive Science at Hampshire College, Amherst, Massachusetts, USA. He is also the director of the Center for the Study of Science in Muslim Societies (SSiMS). His primary research interest focuses on understanding the reception of science in Muslim societies and how Muslims view the relationship between science and religion, as well as how this discourse unfolds in online videos. He teaches 'Evolution, Islam, and Modernity', 'Science in the Muslim World' and 'Science and Religion: Biological Evolution in the Public Sphere' at Hampshire College.

**Jonathan P. Hill** is Associate Professor of sociology at Calvin College. He is author of *Emerging adulthood and faith* (Calvin College Press, 2015) and co-author of *Young Catholic America: Emerging adults in, out of, and gone from the Church* (Oxford, 2014) and *The quest for purpose: The collegiate search for a meaningful life* (SUNY, 2017). He has published articles and book chapters on higher education and religious faith, volunteering, and charitable giving. He also directs the National Study of Religion and Human Origins, a project that explores the social context of beliefs about human origins.

**Stephen H. Jones** is a Lecturer at the University of Birmingham, UK, and serves as current General Secretary of the Muslims in Britain Research Network (MBRN). He is a sociologist of religion specializing in contemporary Muslim communities and public perceptions of science and religion. His published work has focused on, among other things: non-Muslims' perceptions of Islam and science; interactions between lay and professional perspectives on religion and evolution; the political representation of Muslim minorities; and multicultural policies in British local government.

**Tom Kaden** is Academic Council at the Institute of Sociology at Bayreuth University. Previously, he worked as a Research Fellow at York University, Toronto, on the 'Science and Religion: Exploring the Spectrum' project. His research interests in the field of the sociology of religion include creationism, fundamentalism, humanism and atheism. He also worked under the direction of Professor Dr Gerhard Wagner (Goethe University Frankfurt) on Volume 1/7 of the Max Weber Complete Edition. His book *Creationism and anti-creationism in the United States: A sociology of conflict* was published by Springer in 2019.

**Susanne Kind** is a PhD student in the field of Cultural Sociology at the Faculty of Social Sciences and Philosophy, University of Leipzig. She wrote her PhD about the socio-political and ideological activism of members of the Swedish Humanist Association, supervised by Professor Dr Monika Wohlrab-Sahr. From 2013 to 2016, she worked as a research assistant in the Emmy Noether research project 'The Diversity of Non-religion' at the University of Zurich, led by Professor Dr Johannes Quack. Her research interests include humanism, non-religion, religious education, cultural sociology and the sociology of religion.

**Stephen LeDrew** is a Visiting Assistant Professor in the Department of Sociology, Memorial University of Newfoundland, Canada. He is the author of *The evolution of atheism: The politics of a modern movement* (Oxford University Press, 2015).

**Lois Lee** is Senior Research Fellow in Religious Studies at the University of Kent. Her books include *Recognizing the nonreligious: Reimagining the secular* (OUP, 2015), *The Oxford dictionary of atheism* (2016, with Stephen Bullivant) and the co-edited volumes *Secularity and nonreligion* (Routledge, 2015) and *Negotiating religion: Cross-disciplinary perspectives* (Routledge, 2017). She co-directs the Nonreligion and Secularity Research Network (NSRN).

**David E. Long** is Assistant Professor of STEM Education at Morehead State University, Kentucky, USA. Long examines how religious faith and political ideology mediates the US cultural relationship towards evolution, climate science and genetic engineering in educational settings. He is the author of *Evolution and religion in American education: An ethnography* (Springer, 2011).

**Will Mason-Wilkes** received his PhD from Cardiff University in March 2018. His thesis, titled 'Science as religion: science communication and elective modernism', investigated representations of science in non-fiction science television programmes, identifying a 'religious' portrayal of science in some of these programmes, and discussing the potential implications of this kind of representation for public understanding of science, and democracy, in the so-called 'post-truth' age. Will worked as a Research Associate on the 'Science and Religion: Exploring the Spectrum' project at Newman University, Birmingham, and continues to be interested in the intersections of and interactions between science, religion, media, public(s) and democracy.

**Lydia Reid** is a Research Fellow based at St. John's College, Durham University, and is currently working on the project 'Equipping Christian Leadership in an Age of Science' headed by Canon Professor David Wilkinson. Reid has been researching clergy attitudes towards science and is currently focused on their views towards artificial intelligence and technological issues. Prior to this, Reid completed her PhD in Sociology at the University of Manchester and has recently published the monograph *How religious students negotiate the secular culture of a state university* (Edwin Mellen Press, 2017).

**Renny Thomas** is an Assistant Professor in the Department of Sociology, Jesus and Mary College, University of Delhi, India. He was the Charles Wallace Fellow in Social Anthropology at Queen's University Belfast, Northern Ireland, UK (2017–18). He was also the recipient of American Academy of Religion's Collaborative International Research Grant (2015–16), New York, USA. His recent publications include: 'Beyond conflict and complementarity: Science and religion in contemporary India', *Science, Technology and Society*, 23(1): 47–64 (2018); 'Religious rites and scientific communities: *Ayudha Puja* as "culture" at the Indian Institute of Science', *Zygon: Journal of Religion and Science*, 53(1): 95-122 (with Robert M. Geraci, 2018); 'Narratives in feminist sociology of science: Contextualizing the experiences(s) of women scientists in India', in Gita Chadha and M.T Joseph (eds) *Re-imagining sociology in India: Feminist perspectives*, London: Routledge (2018); and 'Atheism and unbelief among Indian scientists: Towards an anthropology of atheism(s)', *Society and Culture in South Asia*, 3(1): 45–67 (2017).

**Amy Unsworth** is a Research Fellow at the University of Kent, UK. She has a long-running interest in science popularization, and more recently her research has focused on how public understandings of science relate to religious and non-religious identities. She has studied perceptions of science, particularly evolutionary theory, among Christians and Muslims using both quantitative and qualitative methods. Previous roles have included developing contemporary science exhibitions and public events at the Science Museum, London, and carrying out research for 'Science, Culture and Modernity', part of a British Council intercultural and inter-religious dialogue project.

# Editors' Acknowledgements

This volume initially arose out of the editors' concern at the lack of scholarly attention given to questions of science and belief within sociology, especially beyond the US. This gave rise to the workshop 'Science, Belief and Society', which was organized by the editors in Birmingham, UK, in February 2016 and sought to bring together sociologists with an interest in this area from across North America, Europe and further afield. We would like to extend our thanks to the University of Birmingham and the Templeton Religious Trust for supporting this event. While the contributors to this volume present research from a range of projects and perspectives, this collection has been developed as part of the 'Science and Religion: Exploring the Spectrum' (SRES) project, and owes a huge debt to that project's Primary Investigator and Co-Primary Investigator, Fern Elsdon-Baker and Bernard Lightman. Our own contributions to this volume draw extensively on Fern's thinking on this subject in particular; many questions we raise in the volume were formulated through conversations with her.

Numerous people have been instrumental in bringing this collection together, and we would like to offer thanks to Alex Hall, Sonja Leuhrmann, Carissa Sharp and Alex Smith for their invaluable advice during the process of developing the collection. We are grateful to Chris Toumey and Thomas Aechtner for comments on some of the chapters (though we should stress that they bear no responsibility for any errors). Will Mason-Wilkes deserves a particular note of thanks for his editorial support in the final stages. Finally, we would like to offer thanks to Victoria Pittman, Shannon Kneis and Alison Shaw at Bristol University Press for taking this project on and for their assistance throughout the process of editing the collection.

# Foreword

*Grace Davie*

This very welcome book explores the interconnections between science, belief and society. In this Foreword, I want to emphasize the urgency of this question (or, more accurately, questions), with reference to a particular example: the publication in 2015 of Pope Francis's second encyclical, '*Laudato Si'*: on care for our common home' (Francis, 2015).

It is clear to all those engaged in the debate surrounding climate change that *Laudato Si'* has been a game-changer, and for two reasons. First, it has vastly extended the range of participants, and, second, it has drawn on established scientific research to deliver a powerful *ethical* message: that deprived communities are likely to suffer most from climate change as currently experienced.

An opinion piece in *The Guardian* puts it thus: 'Pope Francis's encyclical on climate change, *Laudato Si'*, is the most astonishing and perhaps the most ambitious papal document of the past 100 years, since it is addressed not just to Catholics, or Christians, but to everyone on earth' (*The Guardian*, 2015). The following sentence goes further still: it argues that the encyclical sets out a programme for change that is rooted in human needs but makes 'the radical claim that these needs are not primarily greedy and selfish ones' (*The Guardian*, 2015). Rather, we need nature just as we need each other. Thus, the care for nature and the care for the poor are two sides of the same coin. The environment, in the Pope's understanding of the term, captures the underlying connectedness of the world and all those who inhabit it. It is this relationship – the indissoluble link between nature and human beings – that lies at the heart of the document.

Even more striking is the editorial in *Nature*, a prestigious and widely cited scientific journal. Here, the encyclical is warmly commended for the accuracy of the science. However, even more arresting is the awareness that the moral authority and popularity of the Pope will ensure a readership that scientific papers can only dream of; the Catholic Church, moreover, 'has a superb means of spreading the word every Sunday' (*Nature*, 2015). This is not the phraseology that one expects from a scientific journal of this standing.

The final paragraph of the Introduction to *Laudato Si'* articulates the key themes of the encyclical. Relationship and interconnectedness head the list and underpin the subsequent critique. Central to the latter is the impossibility of a technological fix to the problems that we face and the need for new – radically new – ways of thinking. This includes the realization that the ethical questions raised by this debate cannot be resolved by science alone. Finding solutions must become, therefore, an interdisciplinary endeavour.

Unsurprisingly, the encyclical generated huge media interest and a significant secondary literature. Particularly interesting from the point of view of this book is a set of articles, collected under the title, 'The Pope's fateful vision of hope for society and the planet' (Brulle and Antonio, 2015). This includes an editorial and five responses that reflect on the encyclical from a social-scientific perspective. It becomes, in effect, a three-way conversation in which a cluster of social scientists interrogate the relationship between religion, science and social science.

I see *this* volume in a similar light. It offers a further – an instructive – example of the conversations that we need to have about a variety of subjects, ranging from general topics such as science education (Chapter 11) and coverage (Chapter 3), to specific fields such as stem cell research (Chapter 5) and evolution (Chapter 7). For social scientists, this means taking religion seriously as a full partner in the discussion. The challenge, however, is considerable and is best captured in a question: is it possible for a set of disciplines, underpinned as they are by self-consciously secular philosophies of science, to engage with an area of human living that, almost by definition, contests this way of thinking? Precisely this question constitutes the starting point of this volume. Read on.

## References

Brulle, R.J. and Antonio, R.J. (2015) 'The Pope's fateful vision of hope for society and the planet', *Nature Climate Change*, 5(10): 900–1.

Francis (2015) '*Laudato Si'*: on care for our common home'. Available at: http://w2.vatican.va/content/francesco/en/encyclicals/documents/papa-francesco_20150524_enciclica-laudato-si.html (accessed 10 July 2018).

*Nature* (2015) 'Hope from the Pope', *Nature*, 522(7557): 391.

*The Guardian* (2015) 'The Guardian view on *Laudato Si'*: Pope Francis calls for a cultural revolution', *The Guardian*, 18 June. Available at: www.theguardian.com/commentisfree/2015/jun/18/guardian-view-on-laudato-si-pope-francis-cultural-revolution (accessed 6 July 2018).

[illegible]

# Editors' Introduction: Science, Belief and the Sociological Tradition

*Stephen H. Jones, Tom Kaden and Rebecca Catto*

## Science, belief and sociology

Summarizing the sociological study of science and religion a little over 10 years ago, John H. Evans and Michael S. Evans (2008: 88) wrote:

> Although we know of no study of the comparative coherence of sociological research areas, we suspect that the field of religion and science is one of the muddiest in all of sociology. The conceptual source of this muddiness lies in the long-running academic assumption that religion and science always conflict and that they conflict over competing truth claims about the world. It is therefore hard for sociologists to analyze the relationship dispassionately because sociology itself was born as a scientific alternative to religion.

While, as we will see shortly, there have been notable developments since this was written, Evans and Evans's point still stands. Sociological exploration of questions to do with science and religion (or, as we prefer to phrase it, science and belief) is extremely limited and scattered across largely isolated sub-disciplines. Beyond the US (where most research has focused), it would be hard to even describe it as a 'field of study', so limited and disjointed has research been to date. This is surprising in many ways because science and religion has been an abiding subject of public debate for many decades in many parts of the world, and there are numerous ways in which it intersects with sociologists' concerns. Religious voices play a prominent role in many conflicts over science, from stem cell research to, most famously, evolution. There has also been a trend towards religious groups – from evangelicals (Toumey, 1994) to Muslim revivalists (Unsworth, Chapter 12, this volume), to Hindu nationalists (Thomas, Chapter 6, this volume) – justifying their beliefs by making a claim on science (von Stuckrad, 2014). At the same time, science has been deeply entangled in debates about secularization; many non-religious people see science as central to their identity (Lee, Chapter 8, this volume), while humanist and secularist organizations have regularly characterized themselves as 'fighting in the name of science' (Kind, Chapter 9, this volume). Yet, this subject has been on the margins of sociology for many years.

It is hard to understand why this is without, as Evans and Evans say, taking into account the influence of what historians refer to as the 'conflict thesis'. This idea emerged in the 19th century, a period of profound upheaval during which the very idea of the modern 'scientist' emerged,[1] in part due to efforts to establish a sphere of practice for empirical researchers that was independent of clerical authority (Turner, 1978). It is – as Fern Elsdon-Baker and Will Mason-Wilkes explain in Chapter 1 of this volume – an idea most often associated with the work of the American authors John William Draper (2015 [1874]) and Andrew Dickson White (2009 [1896]), but one of the most significant representatives of the conflict thesis is Auguste Comte, the French author who originally coined the term 'sociology'. Comte's (1853) 'law of three stages' posited that societies move from a theological, to a metaphysical, to a 'positive' stage, in which explanations are based on observation, experiment and comparison. Sociology was conceptualized as part of this slow process of replacement, serving, in Comte's (1858) terms, as a 'positive religion' and simultaneously as the 'queen of the sciences'.

Since the 1980s, the notion of 'science' and 'religion' having always existed in a state of conflict has been repeatedly debunked by historians (Brooke, 1991; Lightman, 2001; Numbers, 2010; Harrison, 2015). Within sociology, too, Comte is viewed as a 'museum piece' (Elias, 1984: 33). While his term 'sociology' lives on, his argument that 'positive' knowledge can replace religion is universally dismissed. Even so, the conflict thesis has proved remarkably resilient in the West (Numbers and Hardin, forthcoming; Lee, Chapter 8, this volume). One can perhaps draw an analogy with another term that Comte popularized, 'positivism'. This philosophy – which is rooted in the idea that the procedures of natural science can provide a comprehensive understanding of human actions – has been challenged repeatedly, to the point where, in Anthony Giddens's (1974: ix) words, '[t]he word "positivist" ... has become more of a derogatory epithet than a useful descriptive concept' (see also Gülke, Chapter 5, this volume). Yet, despite this, positivism's underlying idea that societies can be comprehended via scientific procedures still has appeal, meaning that sociology, as George Steinmetz (2005: 3) puts it, 'continue[s] to experience a positivistic haunting'. In the same way, few sociologists openly advocate the conflict thesis; it is rare now to find sociological studies that – to take one case that Evans and Evans (2008: 93) discuss – scientifically test the efficacy of prayer. Yet, it is hard for sociologists to jettison the underlying idea that science is something they *do*, while religion is something they (or some of them, at least) *study*. That being the case, taking on the subject of science and religion together seems to require awkward intellectual contortions.

## The purpose of this collection: boundary crossing

The purpose of this collection is to help consolidate and internationalize the field of science and belief. It aims to bring together the different areas of sociology that touch, albeit lightly, on the subject of science and belief, and to broaden the horizons of the field so that it encompasses a wider range of geographical contexts. These two tasks are subtly interconnected, and require some explanation.

As we mentioned earlier, since Evans and Evans's article was published, there has been a small flourishing of research on science and religion, but this has been largely limited to the US. This work has developed our understanding considerably, and we draw on it here while including two contrasting accounts of science and religion in that country (by Jonathan Hill and David Long). Specifically, US-based sociological research has helped to challenge various conflict thesis-related assumptions about what people believe about scientific theories, and evolution in particular. The conflict narrative generally presumes that *knowledge* is paramount: conflicts between science and religion occur because they are competing philosophical systems whose followers cannot coexist (Kaden et al, 2018). In the last 20 years, however, sociological studies of science and religion focused on the US have moved beyond creationist movements (Barker, 1979; Toumey, 1994) to ask questions such as: are the people who reject evolution actually familiar with creationists' formal arguments (Hill, 2014a; Kaden et al, Chapter 3, this volume)? Does increased knowledge of the science of evolution lead to greater acceptance of it (Evans,

2011; Baker, 2013)? Do religious people think holding correct beliefs about evolution is important (Hill, 2014a), and is it central to religious identity and community (Guhin, 2016)? Is its rejection best predicted by religious commitment, social networks (Hill, 2014b) or the moral positions that a person holds (Evans, 2011; Noy and O'Brien, 2016)?

Alongside this, a body of research has been published that has intervened in a long-standing argument about why scientists in the US are typically less religious than the wider population. Early studies, influenced by the conflict model, contended that this was because of scientists' increased familiarity with scientific knowledge (Leuba, 1934, 2013 [1912]; Larson and Witham, 1998). Elaine Howard Ecklund and colleagues (see, among others, Ecklund and Scheitle, 2007, 2017; Ecklund et al, 2011) have, however, questioned this supposition in a variety of ways, asking questions like: are there alternative explanations for this pattern – such as hostility towards religious people in the sciences (Ecklund, 2010), or the dynamics of professional identity (Ecklund et al, 2008)?[2] Even though scientists tend to be non-religious, do they actually support the conflict thesis (Ecklund and Park, 2009), and those who publicly affirm it (Johnson et al, 2018)? When they do, is this influenced by their scientific work or factors such as the home they were raised in?

We will leave it to other chapters in this collection (notably, Jonathan Hill's, Chapter 2) to unpack some of these questions further. Suffice it to say for now, though, that this body of research, as a whole, has stridently moved away from the conflict thesis, busting popular myths about religious people's beliefs (see, in particular, Ecklund and Scheitle, 2018) and about the coherence and social significance of those beliefs. In this work, identity and moral positioning emerge as arguably more important than knowledge. In morally contested areas of science (and, importantly, *only* morally contested areas of science: Allum et al, 2014; Evans, 2011; Shein et al, 2014), knowledge of scientific theories has an uneven relationship, at best, with acceptance of them. What people profess to accept and reject is often chiefly a function of who they see themselves to be, and whom they mix with, rather than what they understand – and this can be said for scientists (Ecklund et al, 2008) as well as for fundamentalists (Hill, 2014b; Evans, 2018). This argument is most developed in John H. Evans's (2018) recent work, in which he argues that epistemological conflict in the US is best seen as a by-product of a deeper moral conflict. This, in turn, has opened up paths to investigate other, related issues. If only certain (conservative Protestant) groups see evolution as important to who they are (Guhin, 2016), and if even these groups have but a passing awareness of technical arguments about evolution versus creation, then what forces shape public contests between professional creationists and anti-creationists (Evans, 2016; Kaden, 2019)? How do public debates and communal loyalties associated with religion and science influence the everyday settings of American life, especially in that most politically contested of spheres, the classroom (Long, 2011; see also Long, Chapter 11, this volume)?

Out of this body of work, a fuller understanding of science and religion's place in American society has emerged, building on and challenging the simple surveys gauging evolution acceptance that previously dominated empirical research into science and religion in the US (see Hill, Chapter 2, this volume), and that still – with a few notable exceptions (Ecklund et al, 2016; Chan, 2018) – dominate research conducted in other parts of the world (see Carlisle et al, Chapter 7, this volume). In such surveys, people are typically asked to choose from a limited set of evolution belief options, often derived from the formal positions of professional creationist and anti-creationist organizations. Such polling is certainly of value in raising questions, but it tells us little about the wider contexts and meanings ascribed to science and belief. Details of whether, what and why people think about questions of science and belief, and how these perceptions are influenced by moral and cultural conflict, are passed over. Worse still, such polls can often imply that publics have a fixed position on evolution that is directly shaped by professional groups – when, in reality, people's views are more fluid, creative and frequently incoherent and uninformed.

This is a problem that we interrogate in Part I of this book, which explores the strengths and weaknesses of methods previously used in the study of science and belief, and makes proposals to guide future research. After Fern Elsdon-Baker and Will Mason-Wilkes place the sociological discipline in its wider intellectual context in Chapter 1, Jonathan Hill, in Chapter 2, offers a review and critique of US-based survey research on science and religion, pointing to ways in which surveys have led to a distorted picture of Americans' beliefs (exemplified by the misleadingly simplistic but common media claim that close to half of Americans 'believe in creationism': see Barooah, 2012). Then, in Chapter 3, Kaden et al look at whether people see meaning in, or identify with, the labels in terms of which debates about science and religion are conducted ('creationism', 'intelligent design', 'New Atheism', etc). Part I ends with Lydia Reid, in Chapter 4, highlighting the insights that emerge from reflecting on the process and challenges of gathering data related to science and belief. Across the remaining sections of this book the other chapters continue in this vein, presenting research that adopts similar approaches to recent work in the US, but in other contexts – including studies focused on Canada, the UK, India, Sweden, Germany and elsewhere.

## Crossing disciplines

Shifting the geographical context, however, means being forced to look in a slightly different way. With creationist movements being relatively strong in the US, the focus in US-based research has predominantly been on religious people, especially Christian individuals and organizations who reject, and seek to compete with, mainstream scientific institutions. This is not a flaw of individual authors; indeed, there are very good reasons for research in the US to focus on these populations. This concentration of research has, however, engendered some unhappy side effects. It has contributed to creationism in the US being taken as a template for understanding science and religion in other, very different, contexts (see Kaden et al, Chapter 3, this volume). It has also led to a failure to challenge a specific conflict thesis-related stereotype. Many chapters in this book deal with Western European states: places where non-religion is increasingly dominant, the rejection of evolution is more limited (Elsdon-Baker et al, 2017; Unsworth and Voas, 2018) and organized creationism is not influential. In a social-scientific scene dominated by arguments for and against the conflict model, these societies are of little interest: the debate, after all, is over *religion's* impact upon science. However, this, we believe, is intellectually flawed, for it implies an unmediated (and typically positive) link between *non-religious* world views and science.

This goes some way to explaining our preference for using the phrase 'science and belief' rather than the more common 'science and religion'. There is, we suggest, value in studying how moral positioning, cultural norms and legal structures shape interactions between science and non-religious ultimate beliefs, just as there is in the case of religious beliefs. While, of course, this can be done in the US – Ecklund's research with scientists being a good example – turning the focus to Western Europe offers opportunities to examine social contexts where non-religion is more socially dominant and where atheism is not subject to high levels of suspicion (Cox et al, 2015). Moving away from a specific focus on religious people also invites a more in-depth encounter with other areas of sociology than the sociology of religion, where most research in this area is located: specifically, the sociology of science and the (relatively recently emerged) sociology of non-religion. In keeping with this, Part II, Part III and Part IV, respectively, are rooted in the sociology of science, the sociology of non-religion and the sociology of religion. Of course, there is considerable overlap between these, and several of the contributions would fit elsewhere in the book. What we have sought to do, however, is look at what each field brings to this subject and to show readers how these presently isolated fields can speak to one another, as well as how studying science and belief might enrich them.

Each domain has struggled to engage with the subject of science and belief in the past and remains curiously inattentive to the subject. Although there are varied reasons for this, in each case the problem owes something to the way in which the conflict model is in some sense woven into the fabric of sociology, and the conceptual 'muddiness' that this engenders. In the following sections, we outline, in turn, the origins and status of the sociologies of science, non-religion and religion, giving details of the chapters in the relevant sections and explaining how they remedy the neglect of science and belief.

## *The sociology of science*

The history and development of the sociology of science (here used in its broadest sense to encompass the related fields of science and society and science and technology studies [STS]) is too complicated to do justice to here, and is dealt with in greater depth in this collection by Silke Gülke in Chapter 5. Broadly conceived, it is concerned with what Gülke terms the 'social embeddedness' of science, which covers a variety of themes. To follow a metaphor used by one of the field's best-known figures, Thomas F. Gieryn (1999: ix), it can mean looking 'upstream' at the social practices and norms involved in generating certified scientific knowledge (Latour and Woolgar, 1986). It can also mean looking further 'downstream' at what Gieryn (1983, 1999) calls the 'boundary work' that is done in society by various social actors to define and defend the boundaries of what counts as legitimate science – as well as (to extend the metaphor) at the rocks, eddies and gullies of inclusion and exclusion, affecting flow and direction. Finally, it can mean looking towards the 'river mouth' at public perceptions of science or specific scientific theories. Research in this third area typically falls under the heading of the public understanding of science (PUS , or more recently, public engagement in science and technology [PEST]), and, for various historical and institutional reasons (Bauer et al, 2007), is distinct both in its approach and in its underlying normative concerns.

Looking at the development of the sociology of science over time is instructive. Aside from a few notable examples (discussed by Gülke), the early sociology of science tended not to concentrate on the practice of scientific knowledge-making, instead looking at the social function and structure of scientific institutions (Merton, 1979). Increasingly in the late 20th century, however, the sociology of science turned its attention to how knowledge was generated by scientists, and adopted a relativistic approach in which both 'true' and 'false' knowledge was viewed as constructed through social processes shaped by human interests (Brown, 1984; Knorr-Cetina, 1981). In its own way, this struck a blow to the conflict model in sociology by 'levelling' different types of knowledge and applying the same methods to 'scientific' and 'superstitious' ideas. Indeed, some of the pre-eminent figures in the sociology of science have sought to uncover 'myths' associated with scientific rationality and the interests these serve (Latour, 1993, 2011).

As Gülke argues, however, although the sociology of science today involves the treatment of gender, race and class in scientific knowledge production and dissemination, rarely is religion the subject of empirical analysis. This is despite the fact that – as Renny Thomas vividly illustrates in his analysis of data collected with Indian scientists in Chapter 6 – religion can be involved in the process of scientific knowledge-making, just as it can influence research agendas in many countries. This is a less accurate description of PUS: a growing body of sophisticated work is emerging that pays attention to people's beliefs (Allum et al, 2014; Clément, 2015; Elsdon-Baker, 2015; McCain and Kampourakis, 2018; Roos, 2014). In PUS, however, the prevailing normative concern is encouraging publics to engage with and accept science as valid, rather than showing how it is constructed through social processes. This means – as Jessica Carlisle, Salman Hameed and Fern Elsdon-Baker argue in their overview of research into Muslims' perceptions in Chapter 8 – that engagement with questions of ultimate belief typically focus on religious people's opposition to science, with rather clumsy assumptions still sometimes made about these people's 'deficit' in scientific knowledge. Moreover, as Fern Elsdon-Baker and Will Mason-Wilkes explain in Chapter 1, there remains an element of tension between the desire to treat science work critically and PUS's normative aim of promoting a better understanding of and engagement with science. Whatever changes the discipline has been through, vital questions remain.

## *The sociology of non-religion*

The sociology of non-religion offers one of the clearest illustrations of how the conflict model has influenced sociology. In contrast to the other two spheres of enquiry discussed here – which can be traced to the early 20th century and before in the case of the sociology of religion – non-religion, secularity and unbelief were not subjects of sociological enquiry until the very late 1960s (see Bullivant and Lee, 2012). Only really in the last 25 years has non-religion and secularity studies emerged as a distinct field of research. This, as Stephen Bullivant and Lois Lee (2012: 20–1) observe, can only really be explained with reference to anti-religious sentiments:

> Many of the social sciences' early pioneers – Comte, Marx, Durkheim, Freud – themselves avowedly non-religious, were fascinated by religion as the great *explicandum*: how can so many people believe in something so absurd? But in trying to answer this question, and thereby establishing the social-scientific study of religion, they arguably failed to recognize that their own *lack* of belief might itself be amenable to similar research. (Emphasis in original)

While religion has figured in sociology as an exotic and unexplainable thing or an artefact left over from pre-modern times, about to disappear, non-religion has appeared to many sociologists as 'natural' and therefore in need of no explanation. Fortunately, this is a view that has fewer and fewer advocates, but the field of non-religious studies is still very much in its infancy[3] and it has struggled for wider recognition. Despite the fact that 'nones' now count as the world's third-largest belief group (Pew Research Centre, 2015), we have a limited vocabulary to describe varieties of unbelief (Lee, 2015). While research in this field has touched upon the claims that non-religious people make upon and about science (Catto and Eccles, 2013), this work has been given virtually no recognition beyond it: the field's influence on the public understanding of science, for example, is negligible.

In this book, therefore, we include three contributions that explore how non-religious groups and individuals draw upon, and seek to shape the public meanings of, science. In Chapter 9, Lois Lee questions the commonly presumed affinity between non-religion and science, arguing that while one can build a case for this, to be valid, it needs to be grounded in the history of non-religion rather than claims about non-religion being inherently more 'rational'. In Chapter 10, Susanne Kind profiles the Swedish Humanist Association's self-described 'crusade' for science and how the legal structures within which it operates have engendered a split between those who seek to win formal recognition as a community of belief and those who see such steps as reducing science to merely a 'belief among others'. Finally in this section, Stephen LeDrew, in Chapter 10, discusses the place of science in contemporary atheist movements, in particular, examining the 'boundary work' undertaken by New Atheists, whose aim is the legitimation of evolutionary psychology and neuroscience at the expense of the 'relativistic' domain of the social sciences.

## *The sociology of religion*

Religion, of course, has been a central concern for sociologists since the emergence of the discipline in the 19th century, with the three so-called 'founding fathers' of the discipline – Marx, Durkheim and Weber – all including theses on the nature and future of religion as part of their theories of social change. Today, however, the sociology of religion is in a curious position: although a flourishing sub-field, it is in many ways marginal to sociology (Catto, 2015) and suffers from what Jeff Guhin (2014) calls an 'export problem', with its theories and concepts failing to make an impact on the wider discipline. The reason for this is, at least in part, the process of secularization. The sociology of religion in the US and Europe has historically been focused on Christianity, and churches in particular (Cadge et al, 2011; Smilde and May, 2015). As the influence of Christianity has waned, in Europe especially, so the field has seemed of less import. This is compounded by the fact that academic researchers are, as we have already seen, themselves typically non-religious, so it has become an area easy to neglect.

Debates over secularization *theory*, furthermore, have dominated the field. Secularization theory has been central to the sociology of religion in the US as well as Europe (Gorski and Altınordu, 2008; Voas, 2009; Blasi, 2014), with scholars seeking to defend it or develop a competing model, most famously, theories of the 'religious market' (Stark, 1999; Stark and Bainbridge, 1985; Stark and Iannaccone, 1994). More recent theories of secularization (Berger, 1967; Martin, 1978) have tended to eschew, in Bruce's (2002: 26) words, 'a zero-sum view of knowledge', where 'scientific knowledge and rational thought gradually [conquer] territory from superstition', preferring instead to focus on the social functions of religion and how modernization disrupts these (see also

LeDrew, Chapter 10, this volume). Even so, both secularization theory *and* the competing market model have tended to define religion narrowly and substantively, focusing on the social significance of belief in the 'supernatural' (cf Stark and Iannaccone, 1994: 232; Bruce, 2002: 2). This way of defining religion functions to concentrate attention on the vitality (or not) of conventional religious institutions in given social contexts, and not on understanding interactions between different domains of knowledge, practice or discourse. Within these approaches, science remains the practice and religion the object of study, which helps to relegate the analysis of science and religion to the margins of the (marginal) sub-discipline.

In recent decades, however, new approaches have flourished in the sociology of religion that are less interested in documenting the level of religious faith and more interested in mapping, in a theoretically informed way, the changing social formation of belief. This work, which has emerged in tandem with non-religion studies, has sought to show, in Pollack and Pickel's (2007: 604) words, how 'individuals are increasingly freeing themselves from institutional guidelines in their religious ideas and behaviours'. It has turned the spotlight on subjective and everyday lived religion (Ammerman, 2007; Dessing et al, 2013), as well as on the emergence of syncretistic forms of spirituality (Heelas and Woodhead, 2005) and other varieties of 'believing without belonging' (Davie, 1994). These shifts have suggested a variety of ways in which science and religion could be incorporated, in a more theoretically informed way, into the sociology of religion. Such approaches suggest that there is potential value in analysing science and religion as facets of lived identity, for example, and in asking questions about whether people's perceptions of science are becoming syncretic and individualized (Jones and Kaden, forthcoming). What remains curious, however, is that while these developments in the field have made it more fertile for the sociological study of science and religion, few scholars are, so to speak, working the plough. The theoretical barriers have begun to come down, but, outside of the US, sustained attempts to examine these objects of study together remain rare.[4]

This collection includes various contributions that seek to rectify this, from Renny Thomas's analysis of the intersections of class, status and belief in Chapter 6, to Kaden et al's research highlighting the individualized and creative way in which people engage with evolution belief labels in Chapter 3. In Part III, though, we intentionally focus on populations at the heart of debates about science and religion that have nevertheless been sorely neglected in social-scientific analysis. In Chapter 11, David E. Long examines religious 'moderates' in the US, with the focus falling on how those who are religious and nevertheless profess not to have difficulty accepting evolution navigate contexts where conservative evangelicalism is culturally dominant. Then, in the book's final chapter, Amy Unsworth examines the views of British Muslims, concentrating specifically on the influential but neglected 'reformist Islamist' trend.

## Conclusion

Looking back across these three sub-disciplines, one can make out a common transition and set of problems. On the one hand, each has been through a process of change in the late 20th and early 21st centuries, with the different sub-disciplines realigning themselves to take on subjects not traditionally seen as important or even valid subjects of inquiry. In each case, this transition has opened up some space for science and belief to be examined more systematically as a subject in its own right. On the other hand, however, the nascent status of the sociology of non-religion and the isolation of the sociology of science and the sociology of religion from each other (and the broader discipline) means that in practice such systematic analysis has been somewhat limited. What we hope readers will see across this volume are the many ways in which these fields of inquiry can inform each other. Stephen LeDrew's and Susanne Kind's chapters in Part II, for example, are located squarely within the sociology of humanism and atheism, but both describe instructive cases of 'boundary work' that could inform the sociology of science. By the same token, Silke Gülke's chapter's analysis of 'unavailable' knowledge in her ethnography of scientists contains insights that could inform the sociology of non-religious and religious belief alike. It is in these possibilities for disciplinary cross-fertilization and international expansion (which still has much further to go beyond this book) that we see a 'clearer' future for the field emerging.

### Notes

1. The term 'scientist' was coined – by William Whewell – in 1833 and only widely adopted towards the end of the century (Harrison, 2006: 86).
2. This work also includes analysis of how the secularization of scientists varies across academic disciplines, which has been the subject of long-running debate (see Lehman and Shriver, 1968; Thalheimer, 1973; Stark et al, 1996; Ecklund and Scheitle, 2007).
3. The Nonreligion and Secularity Research Network was established only in 2008 (see: https://nsrn.net/).

[4] Until 2018, for example, the topic of science and religion was virtually untouched in perhaps the most innovative UK-based sociology of religion journal, the *Journal of Contemporary Religion*.

## References

Allum, N., Sibley, E., Sturgis, P. and Stoneman, P. (2014) 'Religious beliefs, knowledge about science and attitudes towards medical genetics', *Public Understanding of Science*, 23(7): 833–49.

Ammerman, N.T. (2007) *Everyday religion: Observing modern religious lives*, New York, NY: Oxford University Press.

Baker, J.O. (2013) 'Acceptance of evolution and support for teaching creationism in public schools: the conditional impact of educational attainment', *Journal for the Scientific Study of Religion*, 52(1): 216–28.

Barker, E. (1979) 'In the beginning: the battle of creationist science against evolutionism', *The Sociological Review*, 27(1): 179–200.

Barooah, J. (2012) 'Close to half of Americans believe in creationism according to latest gallup poll'. Available at: www.huffingtonpost.com/2012/06/05/americans-believe-in-creationism_n_1571127.html (accessed 20 October 2015).

Bauer, M.W., Allum, N. and Miller, S. (2007) 'What can we learn from 25 years of PUS survey research? Liberating and expanding the agenda', *Public Understanding of Science*, 16(1): 79–95.

Berger, P.L. (1967) *The sacred canopy: Elements of a sociological theory of religion*, New York, NY: Anchor.

Blasi, A.J. (2014) *Sociology of religion in America: A history of a secular fascination with religion*, Leiden: Brill.

Brooke, J.H. (1991) *Science and religion: Some historical perspectives*, Cambridge: Cambridge University Press.

Brown, J.R. (eds) (1984) 'Scientific rationality: the sociological turn', The Western Ontario Series in Philosophy of Science, Springer Netherlands. Available at: www.springer.com/us/book/9789027718129 (accessed 16 March 2018).

Bruce, S. (2002) *God is dead: Secularization in the West*, Oxford: Blackwell.

Bullivant, S. and Lee, L. (2012) 'Interdisciplinary studies of nonreligion and secularity: the state of the union', *Journal of Contemporary Religion*, 27(1): 19–27.

Cadge, W., Levitt, P. and Smilde, D. (2011) 'De-centering and re-centering: rethinking concepts and methods in the sociological study of religion', *Journal for the Scientific Study of Religion*, 50(3): 437–49.

Catto, R. (2015) 'Sociology of religion in Great Britain: interdisciplinarity and gradual diversification', in A. Blasi and G. Giordan (eds) *Sociologies of religion: National traditions*, Leiden: Brill, pp 107–31.

Catto, R. and Eccles, J. (2013) '(Dis)Believing and belonging: investigating the narratives of young British atheists', *Temenos*, 49(1): 37–63.

Chan, E. (2018) 'Are the religious suspicious of science? Investigating religiosity, religious context, and orientations towards science', *Public Understanding of Science*, 27(8): 967–84.

Clément, P. (2015) 'Muslim teachers' conceptions of evolution in several countries', *Public Understanding of Science*, 24(4): 400–21.

Comte, A. (1853) *The positive philosophy of Auguste Comte*, London: Chapman.

Comte, A. (1858) *The catechism of positive religion*, London: Chapman.

Cox, D., Jones, R.P. and Navarro-Rivera, J. (2015) 'Nonreligious tolerance: American attitudes toward atheists, America's most unpopular religious group', in P.A. Djupe (eds) *Religion and political tolerance in America: Advances in the state of the art*, Philadelphia, PA: Temple University Press, pp 133–50.

Davie, G. (1994) *Religion in Britain since 1945: Believing without belonging*, Oxford: Blackwell.

Dessing, N., Jeldtoft, N., Nielsen, J. and Woodhead, L. (eds) (2013) *Everyday lived Islam in Europe* (new edn), Burlington, VT: Ashgate Publishing Limited.

Draper, J.W. (2015 [1874]) *History of the conflict between religion and science* (1st edn), CreateSpace Independent Publishing Platform.

Ecklund, E.H. (2010) *Science vs religion: What scientists really think*, New York, NY: OUP US.

Ecklund, E.H. and Park, J.Z. (2009) 'Conflict between religion and science among academic scientists?', *Journal for the Scientific Study of Religion*, 48(2): 276–92.

Ecklund, E.H. and Scheitle, C.P. (2007) 'Religion among academic scientists: distinctions, disciplines, and demographics', *Social Problems*, 54(2): 289–307.

Ecklund, E.H. and Scheitle, C.P. (2017) 'The influence of science popularizers on the public's view of religion and science: an experimental assessment', *Public Understanding of Science*, 26(1): 25–39.

Ecklund, E.H. and Scheitle, C.P. (2018) *Religion vs. science: What religious people really think*, New York, NY: Oxford University Press.

Ecklund, E.H., Park, J.Z. and Veliz, P.T. (2008) 'Secularization and religious change among elite scientists', *Social Forces*, 86(4): 1805–39.

Ecklund, E.H., Park, J.Z. and Sorrell, K.L. (2011) 'Scientists negotiate boundaries between religion and science', *Journal for the Scientific Study of Religion*, 50(3): 552–69.

Ecklund, E.H., Johnson, D.R., Scheitle, C.P., Matthews, K.R.W. and Lewis, S.W. (2016) 'Religion among scientists in international context: a new study of scientists in eight regions', *Socius: Sociological Research for a Dynamic World*, 2. Available at: http://journals.sagepub.com/doi/pdf/10.1177/2378023116664353 (accessed 12 April 2018).

Elias, N. (1984) *What is sociology* (new edn), London: Columbia University Press.

Elsdon-Baker, F. (2015) 'Creating creationists: the influence of "issues framing" on our understanding of public perceptions of clash narratives between evolutionary science and belief', *Public Understanding of Science*, 24(4): 422–39.

Elsdon-Baker, F., Leicht, C., Mason-Wilkes, W., Preece, E. and Piggot, L. (2017) *Science and religion: Exploring the spectrum: Summary report of preliminary findings for a survey of public perspectives on evolution and the relationship between evolutionary science and religion*, Birmingham: Newman University and YouGov. Available at: https://sciencereligionspectrum.org/wp-content/uploads/2017/09/SRESYouGov-survey-preliminary-findings-5.9.17.pdf (accessed 15 March 2018).

Evans, J.H. (2011) 'Epistemological and moral conflict between religion and science', *Journal for the Scientific Study of Religion*, 50(4): 707–27.

Evans, J.H. (2018) *Morals not knowledge: Recasting the contemporary US conflict between religion and science*, Oakland, CA: University of California Press.

Evans, J.H. and Evans, M.S. (2008) 'Religion and science: beyond the epistemological conflict narrative', *Annual Review of Sociology*, 34(1): 87–105.

Evans, M.S. (2016) *Seeking good debate: Religion, science, and conflict in American public life*, Oakland, CA: University of California Press.

Giddens, A. (eds) (1974) *Positivism and sociology* (rep edn), London: Ashgate Publishing Limited.

Gieryn, T.F. (1983) 'Boundary-work and the demarcation of science from non-science: strains and interests in professional ideologies of scientists', *American Sociological Review*, 48(6): 781–95.

Gieryn, T.F. (1999) *Cultural boundaries of science: Credibility on the line* (2nd edn), Chicago, IL: University of Chicago Press.

Gorski, P.S. and Altınordu, A. (2008) 'After secularization?', *Annual Review of Sociology*, 34(1): 55–85.

Guhin, J. (2014) 'Religion as site rather than religion as category: on the sociology of religion's export problem', *Sociology of Religion*, 75(4): 579–93.

Guhin, J. (2016) 'Why worry about evolution? Boundaries, practices, and moral salience in Sunni and evangelical high schools', *Sociological Theory*, 34(2): 151–74.

Harrison, P. (2006) '"Science" and "religion": constructing the boundaries', *The Journal of Religion*, 86(1): 81–106.

Harrison, P. (2015) *The territories of science and religion*, Chicago, IL: University of Chicago Press.

Heelas, P. and Woodhead, L. (2005) *The spiritual revolution: Why religion is giving way to spirituality*, London: Blackwell.

Hill, J.P. (2014a) *National study of religion and human origins*, Grand Rapids, MI: BioLogos Foundation.

Hill, J.P. (2014b) 'Rejecting evolution: the role of religion, education, and social networks', *Journal for the Scientific Study of Religion*, 53(3): 575–94.

Johnson, D.R., Ecklund, E.H. and Matthews, K.R.W. (2018) 'Responding to Richard: celebrity and (mis)representation of science', *Public Understanding of Science*, 27(5): 535–49.

Jones, S.H. and Kaden, T. (forthcoming) 'Science and religion as lived experience: narratives of evolution among British and Canadian publics and life scientists', in B. Lightman and F. Elsdon-Baker (eds) *Science and religion: Exploring the spectrum*, Pittsburgh, PA: University of Pittsburgh Press.

Kaden, T. (2019) *Creationism and anti-creationism in the United States: A sociology of conflict*, Cham, Switzerland: Springer.

Kaden, T., Jones, S., Catto, R. and Elsdon-Baker, F. (2018) 'Knowledge as explanandum: disentangling lay and professional perspectives on science and religion', *Studies in Religion/Sciences Religieuses*, 47(4): 500–21.

Knorr-Cetina, K. (1981) *The manufacture of knowledge: Essay on the constructivist and contextual nature of science*, Oxford and New York, NY: Pergamon.

Larson, E.J. and Witham, L. (1998) 'Leading scientists still reject God', *Nature*, 394(6691): 313–13.

Latour, B. (1993) *We have never been modern*, Essex: Prentice Hall.

Latour, B. (2011) *On the modern cult of the factish gods*, London: Duke University Press.

Latour, B. and Woolgar, S. (1986) *Laboratory life: The construction of scientific facts* (2nd edn), Princeton, NJ: Princeton University Press.

Lee, L. (2015) *Recognizing the nonreligious: Reimagining the secular*, New York, NY: Oxford University Press.

Lehman, E.C. and Shriver, D.W. (1968) 'Academic discipline as predictive of faculty religiosity', *Social Forces*, 47(2): 171–82.

Leuba, J. (1934) 'Religious beliefs of American scientists', *Harper's Magazine*, 169: 291–300.

Leuba, J. (2013 [1912]) *A psychological study of religion, its origin, function, and future*, Lenox, MA: HardPress Publishing.

Lightman, B. (2001) 'Victorian sciences and religions: discordant harmonies', in J.H. Brooke, M.J. Osler and J.M.V.D. Meer (eds) *Osiris16: Science in theistic contexts: Cognitive dimensions*, Chicago, IL: University of Chicago Press, pp 343–66.

Long, D.E. (2011) *Evolution and religion in American education: An ethnography*, Dordrecht: Springer.

Martin, D. (1978) *A general theory of secularization*, Agincourt: The Book Society of Canada.

McCain, K. and Kampourakis, K. (2018) 'Which question do polls about evolution and belief really ask, and why does it matter?', *Public Understanding of Science*, 27(1): 2–10.

Merton, R.K. (1979) *The sociology of science: Theoretical and empirical investigations* (new edn), Chicago, IL: University of Chicago Press.

Noy, S. and O'Brien, T. (2016) 'A nation divided: science, religion, and public opinion in the United States', *Socius*, 2: 1–15.

Numbers, R.L. (2010) *Galileo goes to jail and other myths about science and religion* (rep edn), Cambridge, MA: Harvard University Press.

Numbers, R.L. and Hardin, J. (eds) (forthcoming) *'The idea that wouldn't die': The warfare between science and religion*, Baltimore, MD: Johns Hopkins University Press.

Pew Research Centre (2015) *The future of world religions: Population growth projections, 2010–2050*, 2 April, Washington, DC: Pew Research Centre. Available at: www.pewforum.org/2015/04/02/religious-projections-2010-2050/ (accessed 16 March 2018).

Pollack, D. and Pickel, G. (2007) 'Religious individualization or secularization? Testing hypotheses of religious change – the case of Eastern and Western Germany', *The British Journal of Sociology*, 58(4): 602–32.

Roos, J.M. (2014) 'Measuring science or religion? A measurement analysis of the National Science Foundation sponsored science literacy scale 2006–2010', *Public Understanding of Science*, 23(7): 797–813.

Shein, P.P., Li, Y.-Y. and Huang, T.-C. (2014) 'Relationship between scientific knowledge and fortune-telling', *Public Understanding of Science*, 23(7): 780–96.

Smilde, D. and May, M. (2015) 'Causality, normativity, and diversity in 40 years of US sociology of religion: contributions to paradigmatic reflection', *Sociology of Religion*, 76(4): 369–88.

Stark, R. (1999) 'Atheism, faith, and the social scientific study of religion', *Journal of Contemporary Religion*, 14(1): 41–62.

Stark, R. and Bainbridge, W.S. (1985) *The future of religion: Secularization, revival and cult formation*, Berkeley, CA: University of California Press.

Stark, R. and Iannaccone, L.R. (1994) 'A supply-side reinterpretation of the "secularization" of Europe', *Journal for the Scientific Study of Religion*, 33(3): 230–52.

Stark, R., Iannaccone, L.R. and Finke, R. (1996) 'Religion, science, and rationality', *The American Economic Review*, 86(2): 433–7.

Steinmetz, G. (eds) (2005) *The politics of method in the human sciences: Positivism and its epistemological others*, London: Duke University Press.

Thalheimer, F. (1973) 'Religiosity and secularization in the academic professions', *Sociology of Education*, 46(2): 183–202.

Toumey, C.P. (1994) *God's own scientists: Creationists in a secular world* (1st edn), New Brunswick, NJ: Rutgers University Press.

Turner, F.M. (1978) 'The Victorian conflict between science and religion: a professional dimension', *Isis*, 69(3): 356–76.

Unsworth, A. and Voas, D. (2018) 'Attitudes to evolution among Christians, Muslims and the non-religious in Britain: differential effects of religious and educational factors', *Public Understanding of Science*, 27(1): 76–93.

Voas, D. (2009) 'The rise and fall of fuzzy fidelity in Europe', *European Sociological Review*, 25(2): 155–68.

von Stuckrad, K. (2014) *The scientification of religion: An historical study of discursive change, 1800–2000*, Boston, MA: Walter de Gruyter Inc.

White, A.D. (2009 [1896]) *A history of the warfare of science with theology in Christendom*, BiblioLife.

PART I

# Methodological Challenges in the Study of Science and Belief

1

# The Sociological Study of Science and Religion in Context

*Fern Elsdon-Baker and Will Mason-Wilkes*

## Science and religion: a disciplinary blind spot?

The growing interest in the social study of science and religion, to which this volume speaks, is opening up new avenues of qualitative study in an area where there has been a surprising paucity of such research. The chapters in this collection bring together for the first time a range of empirical sites, focusing attention on a variety of religious and non-religious interactions with, and uses of, science. This relative lack of social research on 'science and religion' is surprising given the comparative wealth of research conducted in a variety of disciplines concerning 'religion' and 'science' as separate social and cultural phenomena. This lacuna has led to some issues and problematic assumptions, especially in the way that social-scientific research regarding the relationship between science and religion has framed this relationship (Elsdon-Baker, 2015, 2018). Given the relatively recent emergence of this mode and subject of research, it is both important and timely to engage with the theoretical and methodological backdrop to this growing area of study.

One of the rather unique and exciting facets of this field is that, by its very nature, it draws on expertise and research traditions that do not always ordinarily intersect in the academy. This not only means – as the editors observe in the Introduction to this volume – intellectual traditions within sociology, such as the sociology of religion and the sociology of science and technology, but also those beyond it, such as the history and philosophy of science. This presents some problems that, in part, relate to any multidisciplinary research – significantly, the need to be aware of the norms, traditions, language use and pitfalls to be avoided within each discipline or sub-discipline involved in the research. However, there are also some challenges that are specific to the study of science and religion in society; in this case, the intersection is between two subjects of research that are often perceived to be incommensurable, or, at worst, seen to be in direct and destructive opposition to one another. Thus, any researcher starting out their voyage into what have hitherto been the uncharted waters of the social study of science and religion in society needs to be aware of some of the underlying assumptions that may turn what appears to be plain sailing into a perfect storm of both academic and public criticism. In this chapter, then, we will reflect on some of the cross-disciplinary, historical and philosophical issues that we need to consider in order to peel away some of the core assumptions at play when conducting research into science and religion in their historical, social and cultural contexts.

## Conflict or compatibility narratives

While a more comprehensive research picture of public perceptions, and some of the processes underlying them, is beginning to form, there is still a significant gap in data relating to the perceptions and lived experience of issues around science and religion of wider publics – and, perhaps more importantly, diverse or international publics. To date, a significant proportion of the social science research examining public perceptions of the relationship between science and personal faith, outside of the US, has adopted survey-based or quantitative approaches. While this research has been valuable in serving to highlight the complexity of public attitudes, very little work has been done to understand how these data relate to individual lived experience across different cultural contexts, or the processes by which these attitudes or viewpoints may be formed or reaffirmed. This has inevitably led to some problematic assumptions, especially in relation to the way in which social science research regarding the relationship between science and religion has either overtly or covertly framed this relationship in a binary way as necessarily negative and/or mutually exclusive (Elsdon-Baker, 2015, 2018). On an international scale, this links into broader geopolitical narratives concerning secularization, societal progress and development. Publics' rejection of aspects of science (eg evolution) is still, in some sections of academic or policy-based research, seen to be a measure of either religiosity or a lack of civilizational development (see Carlisle et al, Chapter 7, this volume).

The rather pervasive idea of a necessary clash or conflict between science and religion is referred to by historians of science as the 'conflict thesis'. Its roots are relatively recent (in the historical sense) and can really be traced back to the publication in 1874 of John William Draper's *History of the conflict between religion and science* (Draper, 1874). In the preface to this work, Draper (1874: vi) declared: '[T]he antagonism we thus witness between Religion and Science is the continuation of a struggle that commenced when Christianity began to attain political power'. Draper gives a potted intellectual history from the Greeks onwards in order to evidence that this conflict had its roots in the foundations of 'Roman Christianity' and that the two contending parties had been warring ever since. It is important to note that his main target was not the 'Protestant and Greek churches', which he saw as 'moderate', but rather the Catholic Church, whose leaders he implies are 'extremists'. To Draper, the Protestant principle of individual judgement, if fully embraced, avoided conflict between science and religion. In doing this, he set up what is perhaps now a rather common trope in this discourse that paints science as a value-free enterprise and, conversely, religion (specifically here, Catholicism) as the root of all evil. This trope echo's down the years to the work of perhaps more familiar contemporary commentators like Richard Dawkins[1]:

> As to Science, she has never sought to ally herself to civil power. She has never attempted to throw odium or inflict social ruin on any human being. She has never subjected any one to mental torment, physical torture, least of all to death, for the purpose of upholding or promoting her ideas. She presents herself unstained by cruelties and crimes. But in the Vatican – we have only to recall the Inquisition – the hands that are now raised in appeals to the Most Merciful are crimsoned. They have been steeped in blood! (Draper, 1874: xi)

It is perhaps more understandable for Draper in the late 19th century to take such an uncritical and romanticized view of science as an enterprise. It is, however, less understandable for any modern scholar to adopt such a position, not least given the recognition of the application of scientific discoveries in some of the horrors of the

20th century. It is vitally important to remember that these debates do not occur in a (geo)political or social vacuum. Both Draper and Dawkins, in their respective positing of a conflict between science and religion, have focused on what they perceive to be more extreme forms of religion, in places, tapping into wider social concerns of the day. For Draper, it was Catholicism, and for Dawkins, today, it has increasingly become Islam (see Unsworth, Chapter 12, this volume). It is important, therefore, to understand the historical work undertaken by Draper within its own historical context and the political or social lens through which it was written.

Following on from Draper, in 1896, Andrew Dickson White published *A history of the warfare of science with theology in Christendom*, which is seen as the second canonical text in terms of setting forward a history of perpetual and necessary conflict between Christianity and science (Lightman et al, forthcoming). These two works by Draper and White have set the tenor and tone for a variety of rather 'Whiggish' accounts of a perpetual clash between two warring systems of thought. There are, along the way, commonly referred to examples of what are seen to be key moments of antagonism, from Galileo through to Darwin.[2] Historians of science have rightly challenged the historical validity or perspicacity of Draper's and White's work, as well as subsequent histories or popular tomes that plough a similar furrow (Lightman et al, forthcoming). The recurring myth that, historically at least, there is a necessary conflict between science and religion has been extensively and successfully deconstructed. Most historians who work in this area of study argue instead for what has become known as the 'complexity thesis', most notable among them John Hedley Brooke and Peter Harrison (Brooke, 1991; Harrison, 2015). However, the conflict model has proved very popular and persistent in both popular and scholarly discourse. Uncritical acceptance of this conflict or warfare model has had a profound influence and has no doubt contributed to the idea that science, secularism and atheism go hand in hand. This has also, until recently, tended to pervade the way in which the interaction between science and religion is studied socially, where it has been studied at all.

In the late 20th century, attempts were made to move away from a solely conflict thesis-focused approach and to expand the categorization of positions in regard to the relationship between science and religion. Perhaps, the most well known of these is physicist and scholar of religion Ian Barbour's typology of the possible relationships between science and religion. Barbour famously identified four models or ways in which science and religion might (inter)relate with each other:

conflict, independence, dialogue and integration (Barbour, 1990; see also, and revised, Barbour, 1997):

1. *Conflict*, here, is very much the science *versus* religion (or vice versa) model, or, in Barbour's case, a juxtaposition of scientific materialism (here articulated as a form of scientism) versus biblical literalism. Again, this version of conflict is, in part, introduced through the usual historical instances, for example, the Galileo affair, Darwin and the more recent biblical literalism of modern creationist groups.
2. *Interdependence* is perhaps more widely recognized today as a version of the non-overlapping magisteria (NOMA) model, which was popularized by Stephen Jay Gould (1999). Under this model, science and religion are viewed as independent or autonomous spheres of knowledge, which are largely unproblematic as long as they stick to their own domains. This is sometimes articulated in popular parlance as science answering the 'how' questions and religion answering the 'why' questions – though, of course, demarcating 'how' and 'why' questions is often far trickier than might be assumed.
3. *Dialogue*, for Barbour, covers a range of positions. This could include points at which methodological parallels might exist between science and theology or where their suppositions may be similar. Barbour argues that if science is now understood not to be wholly objective (in the light of the work of Thomas Kuhn: eg Kuhn, 1962), and religion can be viewed to be subject to some paradigmatic behaviours, then there may be methodological parallels. However, he is careful to highlight that the data subject to religious study are vastly different from what we might consider to be scientific data, and it is therefore less possible to test them.
4. *Integration* can, under Barbour's model, be either partial or total, and it implies a commonality of concepts. This could include the re-imagining or reformulation of doctrinal stances to encompass new theories in science, for example, creation doctrine, or it could be a latter-day form of natural theology – whereby evidence for God's word is found by studying God's works, that is, the natural world. The third option given is a fully integrated metaphysical model for which Barbour gives examples from process philosophy, whereby God is both the source of not only order, but also novelty – so God would, in essence, be subject to change over time – a position into which evolutionary thinking can easily be integrated. As Barbour (1990) suggests, this would be a God of 'persuasion' as opposed to a God of 'compulsion'. This final approach may

seem worrisome for science but, conversely, it has fairly significant implications for certain religious positions as it can imply a deity that is non-deterministic, which potentially leads into the furthest realms of deism or pantheism.

This kind of typology might be seen as a welcome antidote to the solely conflict-based approach. However, it can also be seen as an extension of the conflict thesis. As Cantor and Kenny (2001: 770) observe in their critique of Barbour's use of the term 'conflict':

> Once conflict is accepted as governing the terms of engagement, alternative ways of articulating science–religion relations are necessarily skewed by being framed in opposition to conflict. The ease with which the image of conflict could take hold of the historical imagination is in part driven by the use of monolithic categories.

This is a fair critique Barbour's model, which is very much born out of a Christian philosophy-centred – or theology-centred – discourse. What he is really trying to achieve is to map out the intellectual positions of his contemporary scholars. This clearly does not reflect the messy social or historical realities of how people have engaged, or do actually engage, with either science or religion in their day-to-day lives. Cantor and Kenny primarily discuss how Barbour's models are a projection onto historical debates or actors. They conclude that 'only on rare occasions will the individual perceive science-and-religion as a choice between the four essentialist relationships that Barbour postulates' (Cantor and Kenny, 2001: 779). This is not just a historical issue, but an equally important concern for those of us involved in contemporary social data collection. By using these typological categories, are we not from the very outset assuming that conflict is one of the majority positions, if not *the* primary position, held by people and that the other positions are a necessary response to or rejection of conflict, thereby skewing how we approach our data or research dissemination? It goes without saying that any kind of attempt to theoretically categorize abstract positions will, by its very nature, ride roughshod over the complexity or nuances that we might find in lived experience. However, this kind of typological thinking has dominated the representation of the study of science and religion to date. Moreover, while a no doubt useful exercise in mapping out the theoretical landscape, these kinds of categorizations are all too often used to create boxes in which to put people in ways that may often not mean the same thing for them (Elsdon-Baker, 2015). As Reid (Chapter 4, this volume) highlights, where Barbour's categories have been used to guide survey questions, close analysis of responses suggests that they do not map neatly onto the positions of even Christian clergy.

People do not behave or think as neatly as we would often like them to. For example, we have found in our survey-based research that those who endorse a form of necessary clash or conflict between acceptance of evolutionary science and personal faith also endorse a NOMA or interdependence model of science and religion (see also Hill, Chapter 2, this volume). Thus, we can see that NOMA can act as both a compatibility *and* a conflict model.

The call for more qualitative research in this field is, in part, then, a plea to move away from typologies that are based on abstract theoretical models and towards a richer picture of interactions, intersections or lived experiences. It is an unfortunate legacy of debates in this field to date that quantitative research has tended to operationalize these abstract categories in survey instruments, thus applying categorical labels that individuals may not readily identify with (Elsdon-Baker, 2015, 2018; Hill Chapter 2, this volume). Drawing on qualitative findings would, in turn, allow future quantitative researchers seeking to examine societal trends to build a more informed approach to any forms of classification or categorization that are inherently implied by this kind of data collection. Furthermore, rather excitingly, it may also lead to entirely different modes or directions for future quantitative data collection.

Another unwanted side effect of these kinds of typologies is that they actually imply not only a kind of inherent conflict, but also a sort of sliding scale in the way in which they are constructed, as Cantor and Kenny (2001) highlight. To reject conflict between science and religion, it could be implied, means that you must move inexorably towards a kind of synthesis between the two. This certainly plays out in the way that these models are perceived by their opponents, perhaps most vocal among these being Jerry Coyne (eg Coyne, 2015). This small but vocal group see any dismantling of the conflict model as being a sop to religion and a part of a wider (well-funded) hegemonic process. This critique is not just applied to those who are themselves religious and seeking to create a dialogue or some form of integration between their faith and science, it is basically applied to anyone who thinks that religious faith is not highly problematic in light of a scientific world view. Any model of the relationship between science and religion – or, indeed, any scholar – that does not adhere to a strict conflict model is accordingly labelled 'accommodationist'. Being accommodationist is ironically an inclusive club – open to those from all faiths or, like at least one of the authors of this chapter, none. For some of us in the field, being publicly called an accommodationist by Coyne et al is a rite of passage. Our little vessel, attempting to survey uncharted waters, will no doubt get shot at before we have even left harbour – although, of course, another rather sublime irony is that most of the current trends in research that are simplistically being labelled 'accommodationist' by those adopting a sort of extreme scientism position are based on up-to-date empirical evidence. These more recent challenges to the conflict thesis are based on research data which suggest that significant numbers of publics and scientists – across religious and non-religious populations – simply do not view science and religion as part of a strictly or always conflicting narrative. Even a significant proportion of those who reject evolutionary science will happily endorse other areas of scientific research, including genetics (Elsdon-Baker, 2017).

## Epistemic conflict and the deficit model

Alongside the conflict thesis, another related and equally pervasive myth in evidence in this field is, as we saw in the Introduction to this volume, the tacit assumption that this conflict has been and is predominantly, or even solely, epistemic in nature. Again, this is an idea that has been significantly challenged in the past decade by one of the leading scholars in the social-scientific study of science and religion, sociologist John H. Evans (Evans and Evans, 2008; Evans, 2018; see also, among others, Hill, 2014; Ecklund and Schietle, 2018; Noy and O'Brien, 2018). Until very recently, the predominant scholarly gaze in this field was firmly focused on long dead men of science or theology, not publics or, indeed, contemporary scientific actors. This, combined with the assumption of a conflict that was epistemic in nature, led to a rather myopic focus on theological discourses, as opposed to the lived experience of these debates or, indeed, individuals' personal faith.

If scholars have increasingly been challenging these two entwined assumptions – of a trans-historical conflict that is epistemic in nature – why and how, then, does the popular conception that there is a necessary conflict between science acceptance and personal faith persist? The starting point when developing the 'Science and Religion: Exploring the Spectrum' project from which this edited volume grew (see Editors' Acknowledgements) was the premise that locating science and religion in their social context can help us to understand and better answer this question. A good first port of call is to look more widely at public perceptions of science.

Disciplines such as science and technology studies (STS) and scholars interested in the public understanding of science (PUS) have, for a number of years, investigated factors that impact public attitudes to science. Building on pre-existing 19th- and 20th-century social narratives or moral panics concerning a perceived lack of public education in regards to science, a so-called 'deficit model' emerged in the early days of these disciplines (eg Miller, 2001; for additional discussion, see also Carlisle et al, Chapter 7, this volume). Proponents of this model argued, particularly in reference to public attitudes to science, that members of the public were critical of science because they did not possess enough scientific knowledge. Increasing the public's scientific literacy – that is, the stock of knowledge of scientific facts held within the public – it was argued, would increase public support for science (Miller, 1983). Since at least the 1980s, however, this line of thinking has been progressively called into question. Repeated efforts to increase public understanding of science returned negligible increases in public enthusiasm or support for science (Bauer et al, 2007). It is also beginning to be more widely recognized that publics tend not to simplistically reject science wholesale, but rather tend to hold localized concerns relating to single issues of relevance or import to them as individuals, for example, genetic modification, vaccinations or nuclear power. The direct relationship between knowledge and attitudes has thus been called into question. Publics are not simply passive consumers of knowledge. Analyses have shown that on certain issues, more knowledge can, in fact, lead to more negative attitudes (Sturgis and Allum, 2004; Bauer et al, 2007).

What these earlier PUS/science literacy efforts also revealed was the difficulty involved in transmitting scientific or expert knowledge into the public domain. Knowledge does not flow simply and easily from the academy to the public, and the effects of this knowledge when, or if, it eventually arrives in the public consciousness are unpredictable. Thus, it should not come as much of a surprise that we need to look beyond the relative decline of the conflict thesis within certain sub-disciplines or academic fields of study in order to understand its proliferation and continued dominance in the public imagination.

Another problem here, of course, is that while history of science research in this field is far more established, the *social* study of science and religion is, as discussed, still a relatively emergent field of dedicated study. Given the nature of the two foci of study, 'science' and 'religion', there are also a lot of other actors with a stake in the public and scholarly discourse, be they theologians, scientists, religious leaders, professional science communicators/educators or media professionals. We may achieve consensus on a more complex model of engagement between the two foci with colleagues in our field of study. However, any attempts to communicate this more broadly are akin to being in a small fishing boat bobbing around in an armada of warships containing other interested parties with a variety of reasons for inadvertently, or purposefully, promoting conflict (see Kaden et al, Chapter 3, this volume).

Given that a lot of the survey-based research data collected to date in this field has been related to PUS agendas, there has been an unwitting reliance on the idea that knowledge is all that matters when thinking about these issues, which presumes a form of epistemic conflict at its heart. Unfortunately, a deficit model approach is therefore still very much in evidence in research and analysis that has been published in relation to public perceptions about science and religion, despite developments in PUS research that counter it. To clarify, the deficit model plays a role in two ways here: first, in the way in which research is designed or undertaken (an assumption of an epistemic conflict, in part, supposedly arising from a lack of scientific literacy); and, second, in the way in which research findings are utilized (arguing for an increase in public knowledge or education on key scientific issues). A classic (if slightly simplified) example of this would be the underlying assumption in some of the literature that if only we better educated people about evolution, they would no longer be creationists. An approach that is focused on levels of scientific literacy or levels of knowledge ignores all of the contributing social and cultural factors that might lead religious believers to perceive a conflict between the acceptance of evolution and their own personal faith.

## Normative commitments

Assumptions about a public deficit alone are not, however, sufficient to explain why *religion* is so often the central point of discussion. After all, academic interest in, for instance, public understanding of the relationship between science and common sense – which, when considered philosophically, appear to be similarly distinct and potentially opposed knowledge systems – is comparatively less widespread. This philosophical distinction between knowledge systems seems unable to provide, in and of itself, the motivation for the level of scrutiny directed towards the relationship between science and religion in the public imagination. To better understand this, it is helpful to look at the foundational drivers of the social study of science, *and* the social study of religion.

The root of the interest in science literacy and PUS can, in part, be located in a normative belief that science is good for society (Miller, 1983, 2010). The public, this argument goes, needs to understand science so that they can properly operate within, and continue to maintain the functioning of, a modern society. Support for science is a prerequisite if modern 'progressive' society is to exist, and, indeed, the progress of society is to be maintained. In short, this narrative implies that: onward social progress is inherently good; science is required for, and intimately tied up with, progress; and therefore science must be good for society. Interestingly, as Jones et al expand on in the Introduction, this belief was fundamental to the foundation of sociology, too. While it was clearest in the work of Comte, the founding 'triumvirate' of the discipline – Émile Durkheim (2001 [1912]), Max Weber (2002 [1905]) and Karl Marx (1970 [1843]) – all concerned themselves with the decline of religion and the implications of this for society. With the arguable exception of Marx, they did not view the weakening of religious belief and community as unambiguously beneficial. Nor did they see secularization as an epistemic conflict within which scientific knowledge gradually won territory at the expense of the forces of religious irrationalism. Nevertheless, across the early discipline, one of the fundamental assumptions was that, for better or worse, religion and modernity are inherently inversely related. Within the discipline, this is coupled, whether explicitly or implicitly, with Enlightenment assumptions about the relationship between progress, modernity and science – with science seen as a prerequisite of the former two. While offering a critique of some of the potential consequences of modernity, the social theories of Durkheim, Weber and Marx are essentially progressive, in that they see progress of society as inevitable, and arguably as beneficial.

Therefore, connections can be made between the founding aims of sociology and contemporary movements that promote scientific literacy or PUS. If it is assumed that modernity, progress and science are intimately related and intertwined with one another, and, indeed, that without the presence of science, societal progress and modernity are impossible to achieve, then, vice versa, as society progresses and modernity spreads, we would expect to see the greater influence of science. By this argument, the relative scientific knowledge of its citizens is thus seen as a key measure of how modern/progressive a society is. If low levels of scientific knowledge/understanding are present in a society, progress may be stifled, and modernity will be threatened. Therefore, or so the argument goes, scientific literacy must be promoted to ensure progress and modernity. Concurrently, things that hinder scientific literacy are therefore dangerous or problematic as they threaten modernity and progress. Religion, according to this understanding, is opposed to modernity and will be a factor that hinders scientific literacy. It is this view that is very much in evidence within the tired accusations of being an 'accommodationist' thrown at those of us who study the relationship between science and religion, regardless of our own beliefs or world views.

While there are clearly some shared intellectual foundations or underpinning assumptions in the separate disciplinary traditions of the social study of religion and the study of public perceptions of science, we must not mistake that for shared disciplinary norms. Thus, one of the key issues that we are perhaps in danger of overlooking when further developing the study of science and religion in society is the fact that to do so, you have to be cognizant of the different debates and traditions of research/normative values within the study of religion in society and the study of science in society. To put it simply, each area has fundamentally different epistemic and ontological commitments. A sociologist of religion, when conducting fieldwork and analysing their data, does not have to make a truth commitment in relationship to their participants' attitudes or world views. Indeed, there may be points at which they significantly depart from their participants in terms of morality, beliefs and world view. However, ordinarily, it is not an expectation of the study of religion to decide whether or not the existence of a deity/deities is a fact or whether a spiritual system of belief holds water. Furthermore, it is not universally common for sociology of religion researchers to use their platform to evangelize about their own personal faith or world view. Conversely, however, when studying public perceptions of science, there is a commitment to an understanding that, at its core, there is a form of 'truth' about the nature of the universe underlying scientific knowledge, and, regardless of any flaws therein, that the 'scientific method' is by far the best approach to garnering truths about the nature of the universe (of which we are coincidently all a part, including the researchers themselves). So, clearly, research into public perceptions of science has a tendency to view participants' views as either correct or incorrect in terms of a core commitment to scientifically held 'truths' or contemporary consensus positions. So, in both an ontological and epistemological sense, to study science in society or public perceptions of science is to make a commitment (however critical) to a certain model of what the 'true' nature of the universe is or what we might be able to say we *know* about the fundamental nature of the universe. That is not to say that there are not significant and necessary spaces within at least some of those disciplines that investigate the science–society relationship to deconstruct the practice of science or science's role in society, and to query what scientific knowledge actually is or how it might be constructed. Just one of the ways in which this is evidenced is by the voicing of concerns that some science communication, and therefore aspects of science communication research, are at their very heart a part of a giant science public relations (PR) industry. Another example, outlined earlier, is the consensus commitment by researchers in this field to critically tackle the deficit model approach to science communication. At its core, when we are undertaking the social study of anything, we are tacitly accepting the 'truth' that the social world, and therefore the natural world, exists. In the social study of religion or spirituality, by contrast, there is no implicit expectation that researchers should accept that the spiritual world, or, indeed, anything that might be considered beyond the physical world, exists. When we study science in society, that the phenomena that are subject to scientific study are real and that science (including social science) produces reliable (if not certain) knowledge about those social or natural phenomena is, most often, tacitly or explicitly, accepted by researchers.

This can lead to issues when different disciplinary actors become engaged in the social study of science and religion. Someone trained as a sociologist of religion might begin to build models whereby scientific content (and, by extension, the material phenomena subject to scientific study) is classified in an extremely relativistic way as akin to a religious world view or system of belief, an approach that is bound to make many academics who are more used to studying science in society or public perceptions of science more than a little squeamish. The legacy of these differing disciplinary ontological and epistemological commitments for the scholar engaged in the social study of science and religion is that science, on the one hand, cannot be simply treated as one world view among many competing world views, and, on the other, cannot be treated as a wholly objective exercise in truth telling that is disconnected from the social sphere. Our little science and religion fishing boat must navigate not only between warships, then, but also the twin icebergs of social constructivism *ad absurdum* and a sort of hard-line logical positivism.

Conversely, researchers studying science in society or public perceptions of science may also inadvertently or purposefully display a bias against religious or spiritual world views. While there is a (mostly) healthy debate about 'insider' and 'outsider' status in the social study of religion, within the study of science in society, the subject of study or debate – 'empirical science' – is, in places, core to the methodological approaches being employed, which (from either a social-scientific or humanities perspective) are at their heart themselves empirical. It is probably fair to say that a not insignificant proportion of those who work in this area have themselves studied Science, Technology, Engineering, Maths and Medicine (STEMM) subjects earlier in their education or careers. Increasing numbers are actively involved in science communication activities (ourselves included). Some proudly embrace their scientific geek or nerd identity (Again, ourselves included!). Moreover, while STS scholars actually engaging in scientific research in the lab have gained a few raised eyebrows in the past, expressing a love for science or actively promoting it is no longer really seen to be worthy of comment at all, whereas a sociologist of religion converting to and practising the religion they study could well find their credibility questioned. The decades-old 'science wars' perspectives that once pervaded the social theory discussion of science in society have receded. Increasingly, science realism reigns to a degree largely unhindered or unquestioned and a new generation of critically engaged scholars who are broadly supportive of science, warts and all, has replaced the warring factions of years gone by. Science communication and public engagement with science is now, in effect, an industry in its own right, in a way that was not really the case even 20 years ago. In the 1990s, it might have been the case that no serious scientist or researcher would want to be labelled as merely a popularizer of science. Now, entire careers are forged in this territory, and, likewise, there are careers to be made in the social study of it.

## Science and religion research and social identity

That, as researchers, our research questions and methods should be constrained by various social and normative pressures should not come as much of a shock to social scientists, and particularly STS scholars. Ludwig Fleck's (1935) book *The genesis and development of a scientific fact* foregrounds the fundamental role that the social group plays in underpinning individuals' ways of thinking about and understanding the world. Meaningful knowledge and understanding of the world are impossible without a shared social context.

This conceptualization of knowledge and understanding provides a way of making sense of the disciplinary constraints that have acted to demarcate sociological research into science and religion, promoting certain questions and methods and proscribing others. For those social scientists who have studied science and religion, certain aspects of the relationship between the phenomena were, collectively, defined and considered as legitimate to study (resulting in the relationship between the two being, for the most part, operationalized in terms of an epistemic conflict). Certain (overwhelmingly quantitative) methods were considered most appropriate to investigate this relationship (eg measuring scientific literacy), and these ways of carrying out social research into these phenomena formed part of the taken-for-granted reality of being a social scientist in particular geographic locations (particularly the US).

Understanding this also helps us to make sense of the identified disparity between academic and public ways of conceptualizing the relationship between science and religion. It also provides a way for us to make sense of the multiple ways in which this relationship is understood within and by various publics. Public subscriptions to the myriad of positions adopted on the relationship between science and religion can be understood as having their roots in particular social groupings. Understanding public views on science and religion is aided by shifting the focus of research from 'what people know' to 'who people know'.

In certain social groupings, the conflict thesis may be a valid and meaningful way of understanding the relationship between science and religion. Membership of the group both facilitates and requires this kind of understanding. Endorsement of a kind of scientism world view or the conflict thesis can form a part of an individual's or group's social identity. As we have further elucidated in our own research (as part of the wider 'Science and Religion: Exploring the Spectrum' project), it is important to recognize that this kind of commitment to a conflict narrative is a prerequisite to more than one kind of group and extends across religious and non-religious groups or individuals. Looking at online communities – 'New Atheist' groups and message boards on, for instance, Dawkins's website or that of other prominent atheistic evolutionists like Coyne – it is possible to see that membership within these groups requires acceptance of the inevitable and insoluble conflict between science and religion. Similarly, as Kind (Chapter 9, this volume) identifies, contestation between differing interpretations of the correct role for humanism in Swedish society are intimately entangled with different groups' understandings of science. Adherence to either a rejection or acceptance of a certain kind of scientific world view or discrete subject of study (eg evolution or climate change) can and does act as part of our social identity (Jones et al, 2019; Elsdon-Baker, forthcoming; Sharp and Leicht, forthcoming). Or, conversely, it plays a role in our perception of how we expect others to view science based on their social group (Elsdon-Baker, 2015, forthcoming; Jones et al, 2019). It is very important here, however, to delineate between the lived experience of conflict and the perception of conflict in others. So, for example, it may be the case that the *experience of conflict* between an individual's personal faith and a specific scientific area (eg evolutionary science) is actually a minority position in some societies or religious groups (Elsdon-Baker, 2015, 2017). However, it could still be that the *perception of conflict* in others is higher than the actually lived experience of conflict within certain groups (eg humanists) – although this is not to say that a humanist or a religious individual cannot both perceive and experience conflict. As with much of this research, these positions need to be recognized as fluid. Thus, the social group can be pinpointed as key in influencing the beliefs and attitudes of individuals – in shaping how they make sense of the world. What must be foregrounded, however, is the fact that individuals in a complex modern society inhabit multiple social groups at any given time, which provide them with multiple ways of making sense of the world.

The complexity in the formation of individual understandings of the world raises a number of issues regarding traditional research efforts to capture any given individual's understanding. Given the potential instability in attitudes that results from an individual's membership in multiple social groups, methods that seek to capture an individual's fixed attitude on an issue such as science and religion may only prove successful in capturing a momentary snapshot of a much more complex perspective. Problems arise when an aggregate of all these individual snapshots is assumed to provide a coherent and unproblematic picture of the attitudes of the public as a whole. As we have discussed elsewhere (Elsdon-Baker, 2015), when carrying out large-scale surveys of public attitudes towards science and religion, survey design plays a key role in the production of apparent social categories (eg 'creationists') that do not necessarily exist in the social world to the same degree or intensity as the surveys suggest they might. Instead, viewing forms of life as being vital in individual attitude formation, and employing methods that allow the underlying attitudes present in a form of life to be understood through a researcher's long immersion in a group and interaction with its members, allows for a richer and more coherent understanding of attitudes towards science and religion prevalent in society as a whole to be developed.

## Summary

Drawing on the groundwork undertaken by eminent historians of science, we need to likewise move away from an assumption of implicit conflict between science and religion and towards a complexity thesis for the social study of science and religion. We need to be cognizant of the nuance and complexity of how these two concepts interact as part of individuals' lived experience, or as social, group or cultural identities. A key facet of this is also being cognizant of the fact that we, too, are subject to these same social processes and to recognize that our own disciplines, sub-disciplines and fields of study can act as forms of life. Otherwise, some of the nuance in publics' positions, perceptions and interactions with science and religion debates can be lost due to implicit disciplinary biases or, indeed, in some cases, prejudices. We each bring with us from our respective disciplinary track or background certain research norms, views, practices or modes of engagement with our audiences. While this is to a degree the case with any area of multidisciplinary research, it is not ordinarily the case that these are so fundamentally embedded in differing epistemological or ontological positions that can lead the unwitting researcher into intellectually very problematic terrain. We need to employ a form of meta-reflexivity, whereby we are continually mindful of the constraints and taken-for-granted assumptions of our intersecting and respective social science or humanities disciplines. Without this meta-reflexivity, we risk fundamentally failing to ask the right kinds of questions or use the right kinds of methods to understand public attitudes towards science and religion, thus building our implicit biases into the very heart of our research. In part, what we are arguing for here (and elsewhere, see, eg, Elsdon-Baker, 2015, 2018, forthcoming) is that continuing to solely employ less contextual methods or modes of research means that we run the risk of actually distorting that which would be the object of our study, namely, public attitudes towards science and religion. We must avoid our own personal and discipline-specific sirens calling us onto the rocks and always be aware of the two icebergs of social constructivism *ad absurdum* and hard-line logical positivism looming on either side of our little research vessel.

All too often in academic discourse, scholars lose sight of the multilayered and relational ways in which members of a variety of 'publics' relate to 'science', the notion of a scientific society and, indeed, other publics' perceptions, views or stances in relation to 'science'. There is a need to move on from assuming that publics' negative responses to, or lack of endorsement of, aspects of scientific research (eg evolution, climate change, stem cell research, etc) simply transect various epistemological, ontological or ethical narratives. Instead, we need to build more sophisticated and inclusive models of individuals' or differing groups' positions on debates that intrinsically relate to science and religion in society – for example, examining not just 'creationism', but also 'atheistic evolutionism'. This will allow us to better understand why these positions might be expressed, and any implications of this, within the complex systems of geopolitical, cultural and social contexts that lead to individuals' positions on scientific issues acting as an identity marker across a spectrum of religious, spiritual, non-religious and atheistic publics. To do this, we need to look beyond traditional research questions, designs and methods to really get to grips with the complexity of publics' and, indeed, other actors' attitudes towards science and religion. What we can learn from the earlier critiques of the deficit model is that expecting public levels of knowledge, captured in large-scale surveys of individuals, to predict in any neat or orderly way public attitudes towards, or lived experience of, science and religion is at best naive. For this reason, methods that capture a snapshot of an individual's attitudes cannot ever give us the whole picture alone. It is only by combining these findings with those provided when employing methods that seek to provide a richer analysis of the attitudes prevalent in different forms of life that research in this field will provide a clearer picture. There is, in short, a need for more qualitative work that builds on earlier pioneering US-based qualitative fieldwork alongside the more recent and thankfully expanding research in this field. This also needs to include the utilization of more recent quantitative research that is also beginning to challenge some of the old assumptions.

A key problematic in the field to date is that while those of us who work in this area are collectively beginning to recognize the need for, and are indeed building, a more nuanced picture of the multilayered and complex social factors at play at the level of the social group or individual within certain cultural contexts (eg the US and UK) or a broader range of social groupings (eg religious and non-religious groups), until recently, we have neglected to do so with publics within differing cultural settings, be they in different geopolitical settings or with minorities within multicultural settings. This is a far longer voyage for our little science and religion fishing boat, across distinctly choppy theoretical or methodological waters in places, and on which this edited volume is one of the first of a slowly growing fleet to set sail.

## Notes

[1] Indeed, *The root of all evil?* was the title of a television documentary series about religion fronted by Dawkins for Channel 4 on UK television in 2006.

[2] For a good introductory critique of some of the popular myths surrounding both Galileo and Darwin, as well as other rather pervasive historical myths about the relationship between science and religion, see Numbers (2009).

## References

Barbour, I.G. (1990) *Religion in an age of science: The Gifford lectures 1989–1991* (vol 1), London: SCM.

Barbour, I.G. (1997) *Religion and science: Historical and contemporary issues*, San Francisco, CA: Harper San Francisco.

Bauer, M.W., Allum, N. and Miller, S. (2007) 'What can we learn from 25 years of PUS survey research? Liberating and expanding the agenda', *Public Understanding of Science*, 16(1): 79–95.

Brooke, J.H. (1991) *Science and religion: Some historical perspectives*, Cambridge: Cambridge University Press.

Cantor, G. and Kenny, C. (2001) 'Barbour's fourfold way: problems with his taxonomy of science–religion relationships', *Zygon; Journal of Religion and Science*, 36(4): 765–81.

Coyne, J. (2015) *Faith vs. fact: Why science and religion are incompatible*, New York, NY: Penguin Random House LLC.

Draper, J.W. (1874) *History of the conflict between religion and science*, New York, NY, and London: D. Appleton and Company.

Durkheim, E. (2001 [1912]) *Elementary forms of the religious life*, Oxford: Oxford University Press.

Ecklund, E.H. and Scheitle, C.P. (2018) *Religion vs. science: What religious people really think*, Oxford: Oxford University Press.

Elsdon-Baker, F. (2015) 'Creating creationists: the influence of "issues framing" on our understanding of public perceptions of clash narratives between evolutionary science and belief', *Public Understanding of Science*, 24(4): 422–39.

Elsdon-Baker, F. (2017) 'Questioning evolution is neither science denial nor the preserve of creationists', *The Guardian*, 5 September.

Elsdon-Baker, F. (2018) 'Re-examining "creationist" monsters in the uncharted waters of the social study of science and religion', in B. Nerlich, S. Hartley, S. Raman and A. Smith (eds) *Science and the politics of openness: Here be monsters*, Manchester: Manchester University Press, pp 258–77.

Elsdon-Baker, F. (forthcoming) 'Creating hardline "secular" evolutionists: the influence of question design on our understanding of public perceptions of clash narratives between evolutionary science and belief', in F. Elsdon-Baker and B. Lightman (eds) *Science and religion: Exploring the spectrum*, Pittsburgh, PA: University of Pittsburgh Press.

Evans, J. (2018) *Morals not knowledge: Recasting the contemporary U.S. conflict between religion and science*, Berkeley, CA: University of California Press.

Evans, J.H. and Evans, M.S. (2008) 'Religion and science: beyond the epistemological conflict narrative', *Annual Review of Sociology*, 34: 87–105.

Fleck, L. (1935) *The genesis and development of a scientific fact*, Chicago, IL: University of Chicago Press.

Gould, S.J. (1999) *Rock of ages: Science and religion in the fullness of life*, New York, NY: Penguin Random House LLC.

Harrison, P. (2015) *The territories of science and religion*, Chicago, IL: University of Chicago Press.

Hill, J.P. (2014) 'Rejecting evolution: the role of religion, education, and social networks', *Journal for the Scientific Study of Religion*, 53(3): 575–94.

Jones, S., Catto, R., Kaden, T. and Elsdon-Baker, F. (2019) '"That is how Muslims are required to view the world": race, culture and belief in non-Muslims' descriptions of Islam and science', *The Sociological Review*, 67(1): 161–77.

Kuhn, T. (1962) *The structure of scientific revolutions*, Chicago, IL: University of Chicago Press.

Lightman, B., Nickerson, S. and Tajbakhsh, P. (forthcoming) 'From conflict to complexity: historians and 19th century public perceptions of science and religion', in F. Elsdon-Baker and B. Lightman (eds) *Science and religion: Exploring the spectrum*, Pittsburgh, PA: University of Pittsburgh Press.

Marx, K. (1970 [1843]) 'Critique of Hegel's philosophy of right', in J. O'Malley (ed) *Marx: Early political writings*, Cambridge: Cambridge University Press, pp 1–27.

Miller, J.D. (1983) 'Scientific literacy: a conceptual and empirical review', *Daedalus*, 112(2): 29–48.

Miller, J.D. (2010) 'The conceptualisation and measurement of civic scientific literacy for the twenty-first century', in J. Meinwald and J.G. Hildebrand (eds) *Science and the educated American: A core component of liberal education*, Cambridge, MA: American Academy of Arts and Sciences, pp 241–55.

Miller, S. (2001) 'Public understanding of science at the crossroads', *Public Understanding of Science*, 10(1): 115–20.

Noy, S. and O'Brien, T.L. (2018) 'An intersectional analysis of perspectives on science and religion in the United States', *The Sociological Quarterly*, 59(1): 40–61.

Numbers, R. (ed) (2009) *Galileo goes to jail and other myths about science and religion*, Cambridge: Harvard University Press.

Sharp, S. and Leicht, C. (forthcoming) 'Beyond belief systems: promoting a social identity approach to the study of science and religion', in F. Elsdon-Baker and B. Lightman (eds) *Science and religion: Exploring the spectrum*, Pittsburgh, PA: University of Pittsburgh Press.

Sturgis, P. and Allum, N. (2004) 'Science in society: re-evaluating the deficit model of public attitudes', *Public Understanding of Science*, 13(1): 55–74.

Weber, M. (2002 [1905]) *The protestant ethic and the spirit of capitalism*, New York, NY: Penguin.

White, A.D. (2009 [1896]) *A history of the warfare of science with theology in Christendom*, Cambridge: Cambridge University Press.

2

# Survey-based Research on Science and Religion: A Review and Critique

*Jonathan P. Hill*

## Introduction

A renewed theoretical and empirical programme at the intersection of science and belief must begin by taking stock of our present resources. This chapter provides a top-level overview of the survey-based empirical data currently available. The goal is not to provide a traditional 'literature review' of existing studies, but rather to give an assessment of the data themselves – their substantive focus, promise and limitations. Moreover, this is not an exhaustive, encyclopaedic account of every survey measure related to science and religion. That would surely require its own volume. Instead, I focus on direct attempts to measure public attitudes about the relationship between science and religion. Practically, this means that I limit myself to measures of the so-called 'conflict thesis' (see the Introduction to this volume) and beliefs about evolution and human origins. I also limit my analyses to probability samples of the general population in the US (where the overwhelming bulk of this research has been conducted). Despite these caveats and boundaries, several important features of and limitations to the existing data emerge from this review. In short, I find that certain measures work well at a basic descriptive level but many of the important questions about *why* segments of the population hold to some positions and not others may require new measures and research strategies. The chapter concludes with some specific suggestions for what forms these new directions might take, many of which are taken up in various ways in the remaining contributions to this volume.

## Survey items on the conflict thesis

In 1957, the National Association of Science Writers (NASW) sponsored a national poll on the American public's interest in science and technology (Davis, 1957). The Institute for Social Research at the University of Michigan administered it just a few months before Sputnik I launched. The survey instrument that was developed formed the basis for numerous public opinion questions that were later replicated in the US and around the world. In 1978, after a period of relative silence on these topics in national surveys, the National Science Foundation (NSF) recruited the political scientists Jon Miller and Kenneth Prewitt to design a national survey to follow up the original 1957 study (Miller et al, 1980). The NSF's Surveys on Public Understanding of Science and Technology were regularly administered between 1979 and 2001.

The operating framework for these early studies was, as others have termed it, a 'deficit' model of public understanding and attitudes (Dickson, 2005). The driving concern was that segments of the public were either not knowledgeable enough about science or lacked the positive attitudes towards science that elites desired (Bauer et al, 2007). While religion was not a primary focus of these early surveys, when it did arise, it was frequently framed as a potential cause of this deficit or a potential obstacle to overcoming it.

The longest-standing survey question that captures conflict between science and religion comes from the 1957 NASW survey. The item asks participants to agree or disagree with the following statement: 'We trust too much in science and not enough in faith'. The question has become something of a staple in surveys on science and was replicated in the NSF-sponsored surveys throughout the 1980s and 1990s and included in the General Social Survey (GSS) and numerous other national and international surveys. Figure 2.1 shows the percentage of the US population that agrees or strongly agrees with this statement over time. Although it tends to hover around 50% for most of the latter half of the 20th century, there does appear to be a slight decline in agreement over time (with the most recent survey in 2010 showing 42% agreeing). However, if this question is designed to pinpoint a position on the relationship between science and religion, it does so poorly. Individuals who both strongly agree or strongly disagree

**Figure 2.1: Percentage agreeing with the statement: 'We trust too much in science and not enough in faith'**

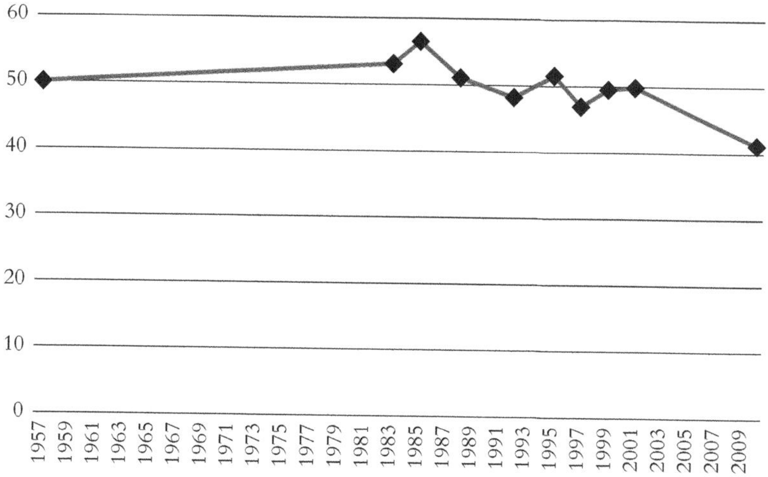


*Sources:* The 1957 NASW survey, the 1983–2001 NSF Surveys of Public Understanding of Science and Technology, and the 2010 GSS

could *both* quite plausibly be committed to a belief that science and religion conflict.

In recent years, some surveys have been including questions that ask about belief in conflict in a more direct way (though we will soon see that even this is difficult to interpret). The most basic survey question about this indicates that a slight majority of Americans believe that science and religion conflict. Surveys that ask respondents to indicate whether they believe that science and religion are 'often in conflict' or 'mostly compatible' are fairly consistent: slightly over half choose 'often in conflict' (54–55%), 38–40% choose 'mostly compatible' and 5–7% are unsure what they believe (see Figure 2.2).

On their face, these data seem to suggest that a little more than half of the American public believes that science and religion are in opposing camps: one favours reason and empirical evidence; the other faith and divine revelation. In other words, this looks like evidence for belief in the conflict thesis. However, this survey question, probably more than most, is sensitive to wording and framing. Variations on this question produce quite different results. If, instead of two possible positions, the respondent is asked to pick from *three* possible positions – that science and religion 'generally agree', 'generally conflict' or are 'not related to each other in any meaningful way' – then only between one quarter and one third pick the conflict response (see

**Figure 2.2: In your opinion, generally do you think science and religion are often in conflict or science and religion are mostly compatible?**

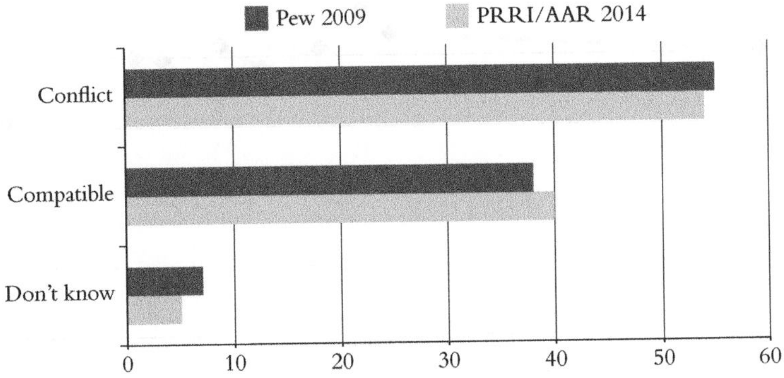


*Sources:* The 2009 Pew Research Center for the People and the Press/The American Association for the Advancement of Science Survey and the September 2014 Public Religion Research Institute and American Academy of Religion (PRRI/AAR) Survey on Religion, Values and Climate Change

Figure 2.3). This additional option – that the two are independent from one another – corresponds closely to the late Stephen Jay Gould's (1997) concept of non-overlapping magisteria (NOMA). This view holds that science's proper domain is the factual description of the world, while religion's is the realm of ethics and ultimate values. A little over one third of the US population (36–37%) seem to agree.

For Gould, this was a way to *avoid* conflict between religion and science, but this is not the perception of many in the American public (on the Canadian and UK publics, see also Elsdon-Baker and Mason-Wilkes, Chapter 1, this volume). Although the addition of a third option reduces respondents from both of the other positions, the reductions from the 'conflict' position are greater. The association between the NOMA position and the conflict position is also clearer in the National Study of Religion and Human Origins (NSRHO), a national survey of the US public that I conducted in 2013 (Hill, 2014a). This survey includes an item that asks respondents to rate their agreement with the following statement: 'Science is about facts and religion is about faith. The two do not overlap'. One third of the population agrees with this statement, quite close to the proportion affirming the independence position in other surveys. Those who agree with this statement are actually *more* likely to agree with another statement: 'Sometimes I feel like the findings of science and the teachings of [my] religion conflict with each other' (68% versus 48%

**Figure 2.3: Which comes closer to your view about the relationship between science and religion? They generally agree with each other, they generally conflict with each other, or they are not related to each other in any meaningful way?**

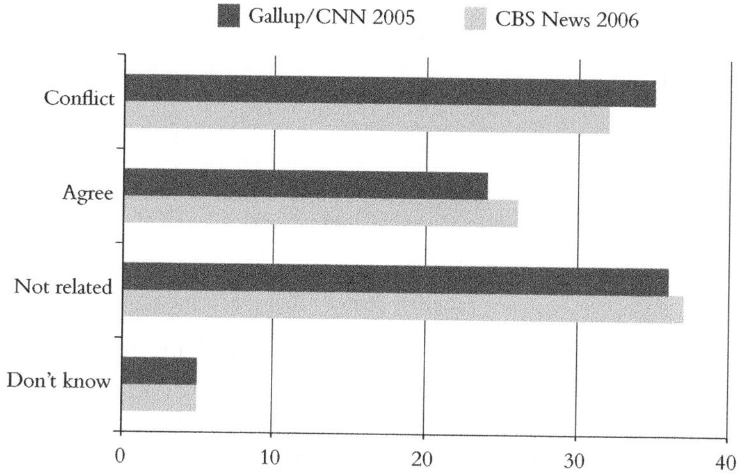


*Sources:* CNN, conducted by Gallup, 2005, and CBS News, 2006

for those that disagreed with the NOMA position). Furthermore, they were *less* likely to agree with the following statement: 'Ultimately, I believe that the findings of science and the teachings of [my] religion are compatible' (27% versus 50% of those who disagreed with NOMA).

Still, many who hold to the conflict position end up rejecting NOMA. Do they believe that science and religion are fundamentally incompatible and competing ways of gaining knowledge about the world? Some probably do, but it is unlikely that all do. The 2007 Baylor Religion Survey included a slightly different question that helps to illustrate this. Instead of asking whether science and religion conflict, they asked whether science and religion were incompatible. Only 17% of Americans agreed with this statement (see Figure 2.4). This is considerably less than the one quarter to one third who identify with the conflict position in the three-position question. In other words, there is a group of people who believe that science and religion come into conflict but probably do not hold that science and religion are *inherently* at odds with one another.

In many ways, this 17% fits most closely with a strong epistemological warfare position. This view holds that science and religion are rival systems for making claims about the world, and one must choose which side is more reliable. The evidence suggests that this group

**Figure 2.4: Please indicate your level of agreement: science and religion are incompatible**

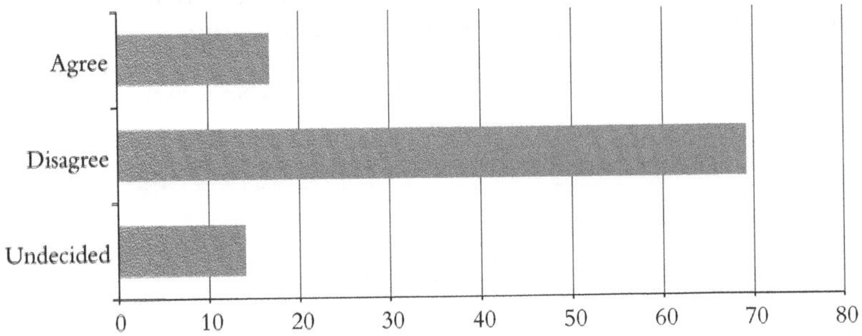


*Source:* The 2007 Baylor Religion Survey

is split evenly between siding with religion and siding with science (Scheitle, 2011; Baker, 2012).

Let us return to those who agree that the two are in conflict but are not fundamentally incompatible. What might they be thinking about when they answer the conflict question? For many, the answer is likely evolution and creationism. A 2009 Pew survey on science asked respondents who felt that science conflicted with their own religious beliefs to report what the conflict was about. Evolution and Darwinism were cited more than four times as much as the next most common answer (which was stem cells, followed closely by abortion) (Masci, 2009). For many, then, the relationship between religion and science is probably not one of general epistemological conflict, but one of conflict over very particular knowledge claims. The sociologist John Evans (2011) finds something similar among conservative Protestants. Compared to other religious and non-religious groups, conservative Protestants were just as likely to seek out scientific knowledge (for similar findings, see Scheitle et al, 2017), something that would be unusual if they were operating from a framework of general epistemological conflict. At the same time, Evans found that conservative Protestants were more likely to register opposition to evolution and the Big Bang. While these specific conflicts are important to many Americans, they do not necessarily lead to a belief in the general incompatibility of science and religion.

To summarize, although a slight majority indicates that they believe science and religion are 'often in conflict', the meaning behind this claim is not uniform. If we take this group of 55–60% of Americans, we can further divide them into three additional groups. When given the additional option, roughly one third reports that science and religion are completely separate domains. The most probable explanation is that they are likely to *see* prominent public debates about religious and scientific claims, but probably believe that these are ultimately unnecessary conflicts because one or the other is stepping out of its natural domain. Another third of the conflict group appear to believe that there are real conflicts between science and religion but that these conflicts are limited to very specific domains and not to a general epistemological conflict. The last third of those that report conflict probably *do* view the practices of religion and science to be fundamentally 'incompatible'. Within this group, about half side with religion and half with science.

All of this suggests that there is more than meets the eye to survey questions about the relationship between science and religion. Although the results appear to have a straightforward reading at first, it is important to recognize that changes in the wording and framing of these questions are tapping into different conceptions of science, religion and the boundaries between the two. Towards the end of this chapter, I revisit these findings and suggest some alternative ways to measure belief in the conflict thesis.

## Survey items about human origins

Survey questions about human origins are more common in the US than questions about religion and science conflict. Most can be classified into two types: (1) direct questions about belief in human evolution; or (2) questions that attempt to have respondents self-classify as evolutionists or creationists.[1] There are a few exceptions to these – along with an entirely different set of questions that attempt to measure what should be taught in schools – but these are the primary strategies for measuring what individuals themselves believe.

A standard version of the first type asks respondents to indicate whether they think that the following statement is true or false: 'Human beings, as we know them today, evolved from earlier species of animals'. In most versions of this, the American public is very close to evenly split, with a slight edge for the 'true' option. Figure 2.5 shows the results from this question in various surveys between 1985 and 2014. Viewing the data longitudinally naturally brings up the question of whether there has been historical change in public opinion. The results seem to indicate very few discernible trends over time.

As with survey items on the conflict thesis, variations on this question can help us glean additional clues about what the public believe. Instead of simply asking whether the statement is true or false,

**Figure 2.5: Human beings, as we know them today, developed from earlier species of animals**

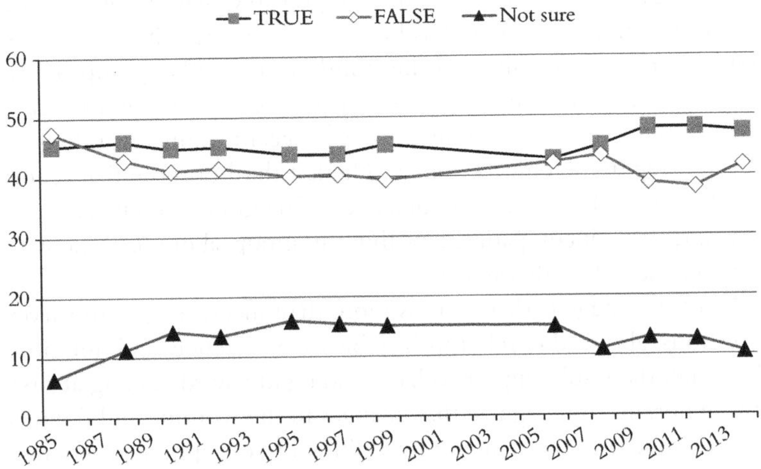


*Sources:* The 1985–99 NSF Surveys of Public Understanding of Science and Technology and the 2006–14 GSS

respondents in one variation are asked whether it is 'definitely true', 'probably true', 'probably false' or 'definitely false'.[2] In all instances of this question, respondents are considerably more likely to select 'definitely false' than 'definitely true' (see Figure 2.6). Although a slightly greater proportion believes that human beings evolved, this group seems to be less confident of their position than the group that denies human evolution. Looking over time, however, this confidence gap appears to be shrinking.

Another variation allows respondents to indicate that they believe 'both' the account of evolution given by scientists and the biblical account. This question has only been included in surveys twice, once in 1999 and once in 2011. The results can be found in Figure 2.7. Two features are worth noting. First, a little more than a quarter of the respondents in both surveys choose the 'both' option. In other words, they do not see a necessary contradiction between the biblical account and Darwinian evolution. Second, the size of the group who chooses evolution appears to be growing larger (from 15% to 21%) and the group choosing the biblical account is receding in size (from 50% to 45%). While we should be cautious in assuming trends with only two points of data, it is notable that this trend is consistent with the last figure that suggests similar modest shifts in belief over time.

This shift over time is also consistent with the last variation of this question that I will cover. This is the variation that is preferred in

**Figure 2.6: Human beings developed from earlier species of animals**

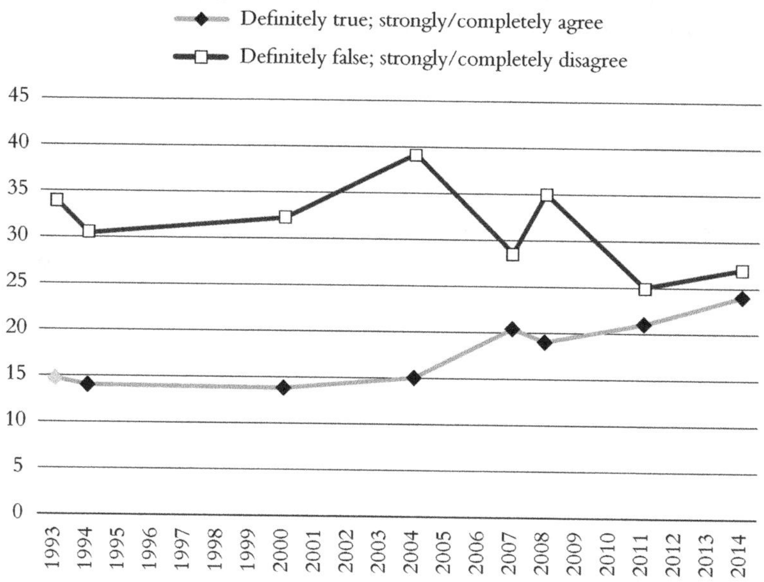


*Sources:* The 1993–2004 GSS; the 2007 Pew Forum on Religion and Public Life US Religious Landscape Survey; the September 2008 Yale University/George Mason University Global Warming's Six Americas Survey; the September 2011 CNN/ORC International Poll; and the September 2014 PPRI/AAR Religion, Values and Climate Change Survey.

surveys conducted by the Pew Research Center. The question asks respondents which view comes closest to their own: humans evolved over time or humans have existed in their present form since the beginning of time. This question (see Figure 2.8) has been asked in a variety of Pew surveys since 2005. The results tell us that evolution is the favoured response overall, and that this response is growing modestly in popularity. Around half the population chose evolution in the mid-2000s; in the most recent years, this estimate has been 60% or more.

We might wonder why there are differences in trends over time between the item that Pew uses and the item that the NSF surveys and the GSS uses. There is no way to know for sure without including both questions in the same survey (which has never been done); however, we can speculate about probable reasons. The most notable difference between the two questions is that the former specifically links humans to non-human ancestors while the Pew question simply acknowledges that human beings have 'evolved'. Some Americans could hold to a

**Figure 2.7: Which do you think is more likely to actually be the explanation for the origin of human life on Earth? The theory of evolution as outlined by Darwin and other scientists, the Biblical account of creation as told in the Bible, or are both true?**

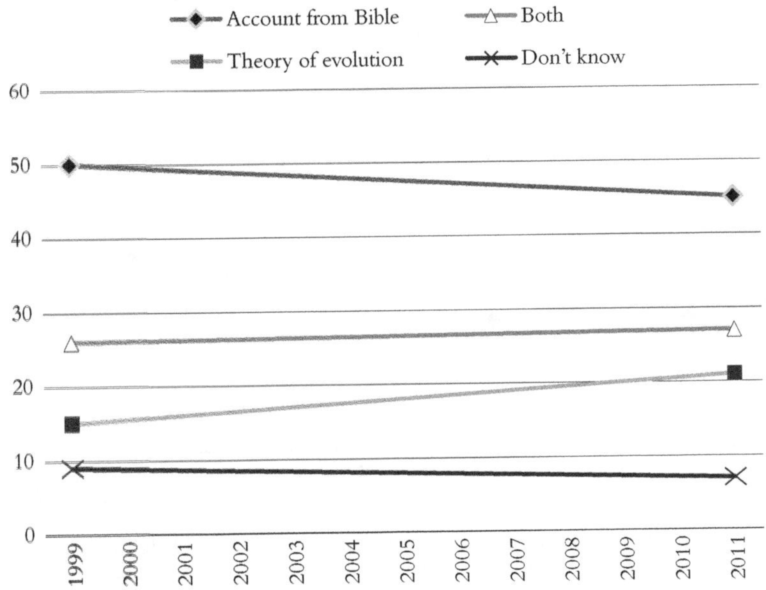


*Source:* Fox News, 1999 and 2011

view of evolution that believes humans are changing and adapting over time, without necessarily believing in the transmutation of the species. Or, this difference could simply be the result of a visceral reaction against explicitly linking humans to non-humans in the former question. In any case, there appears to be more acceptance of evolution when the question is framed in the way that Pew chooses to.

This leads us to the second type of question: the self-identity question. This question was made popular by the Gallup organization. Respondents are asked to sort themselves into categories that roughly correspond to a theistic evolutionist position, a non-theistic evolutionist position or a Young Earth creationist position. Figure 2.9 shows the breakdown of this question between 1982 and 2017. The Young Earth creationists appear relatively stable over time – somewhere between 38% and 47% of the population. There is the possibility of a slight downward dip in recent years but a few additional surveys are probably needed to confirm this trend. The most noticeable trend is the increase in the non-theistic evolutionist position, moving from 9% of the population in 1982 to 19% in 2017. The theistic evolutionist

**Figure 2.8: Which comes closer to your view? (a) Humans and other living things have evolved over time, or (b) humans and other living things have existed in their present form since the beginning of time**

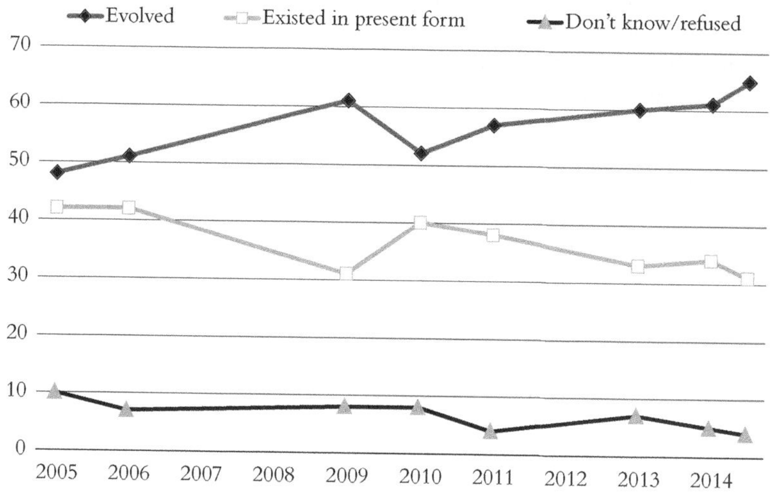


*Sources:* The 2005 and 2006 Pew Forum on Religion and Public Life; the 2009 Pew Research Center for the People and the Press/The American Association for the Advancement of Science Survey; the 2010 Pew Forum on Religion and Public Life Religious Knowledge Survey; the 2011 PRRI/RNS Religion News Survey; the 2013 Pew Research Center for the People and the Press Poll; the 2014 Pew Research Center for the People and the Press Political Typology Survey; and the 2014 Pew Research Center General Public Science Survey

group appears to be mostly stable over time, with between 31% and 38% adopting this view.

The Gallup question is useful to gauge where people place themselves, but it has come under some fire in recent years. One criticism is that it does not accurately cover the possible spectrum of views on human origins. Most notably, there is nothing like an Old Earth creationist view that posits the special creation of humans by God within an ancient geological time frame. In fact, those who want to affirm some aspect of human evolution but also desire a strong role for divine intervention will likely feel conflicted over whether to choose the Young Earth creationist position or the theistic evolutionist position.

In an unpublished presentation (Higgins et al, 2011), researchers included both the Gallup survey question (three categories) and a new four-category question on human origins: (1) God created all life in its current form and no evolution has taken place; (2) God both created and guided evolution; (3) God created but evolution occurred

**Figure 2.9: Which of the following statements comes closest to your views on the origin and development of human beings? (a) Human beings have developed over millions of years from less advanced life forms, but God guided this process, (b) human beings have developed over millions of years from less advanced life forms, but God had no part in this process, or (c) God created human beings pretty much in their present form at one time within the last 10,000 years or so**

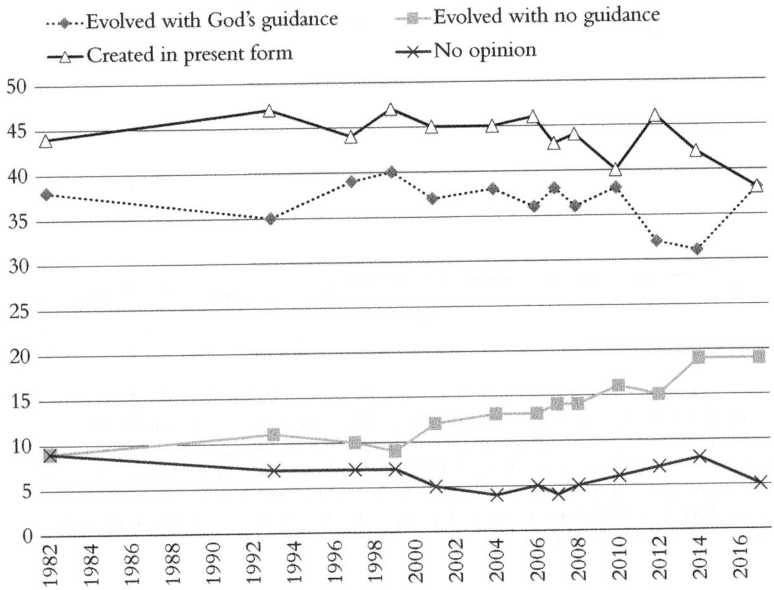


*Sources:* Gallup, 1982, 1993, 1997, 1999, 2001, 2004, 2006, 2007, 2008, 2010, 2012, 2014, 2017

naturally; and (4) human beings have developed over millions of years from far less advanced life forms, but God had NO PART in this process. Over 13,000 respondents answered both questions. The cross-tabulation of these results is presented in Table 2.1. Only 36% of those who select the Young Earth creationism option in the Gallup question also select the full creationism option in the four-category alternative. Around half of this group selects the God-guided evolution option instead. These results make it clear that when people are selecting the Young Earth creationism option in the Gallup surveys, they are selecting this as what intuitively comes closest to their beliefs given the options presented. We should not make the mistake of assuming that they are all fully fledged, firmly committed, Young Earth creationists.

**Table 2.1: Survey experiment results for alternative to Gallup measure on human origins**

| | Four-category measure (%) | | | | | |
|---|---|---|---|---|---|---|
| **Gallup measure** | **Creationism** | **God-guided evolution** | **God origin/ evolution independent** | **Evolution** | **None** | **Not at all sure** |
| Creationism | 36.4 | 49.6 | 7.1 | 0.6 | 2.6 | 3.7 |
| God-guided evolution | 4.1 | 62.8 | 17.8 | 3.2 | 5.1 | 7.0 |
| Evolution | 0.7 | 0.9 | 7.6 | 82.0 | 5.4 | 3.3 |
| None comes close to my beliefs | 10.4 | 23.0 | 11.2 | 7.9 | 38.1 | 9.4 |
| Not at all sure | 4.3 | 14.3 | 8.9 | 5.5 | 7.5 | 59.5 |

*Source*: Higgins et al (2011)

To summarize, what can we conclude from the survey items on human origins? First, a slight majority of Americans accept that humans have evolved. Second, this group appears to be less confident in their beliefs than anti-evolutionists. Third, a majority of those who accept evolution do *not* believe this to be opposed to the biblical account. Fourth, the percentage of the population that accepts evolution, is confident in this belief and does *not* believe that this is compatible with the biblical account is growing slightly over time. Lastly, self-identity question types (particularly the three-category Gallup question) are sensitive to the available categories provided and therefore probably do a poor job at accurately representing the public's beliefs about human origins.

## Further analysis

Basic descriptive statistics only get us so far in understanding public perceptions. We can venture a bit further by assessing how certain religious factors help to differentiate beliefs about science and religion. Do some of these beliefs fall neatly along the lines of religious identity, practice and belief while others do not? In order to investigate these questions, I ran statistical analyses that measured the impact of some of the most common religious measures used by sociologists: a measure of religious tradition, a measure of Bible beliefs and a measure of religious service attendance. For each measure, I ran bivariate and multivariate logistic regressions. The results are presented graphically in Figures 2.10–2.17. The multivariate models (represented by the 'adjusted' values) typically control for gender, race/ethnicity, age, income, educational attainment, employment status, marital status, political ideology and region of the country. This method eliminates potential spurious factors that may be influencing the relationship between the religion measure and the outcome being examined. The figures also include error bars that indicate the 95% confident interval of the point estimate. These help to determine whether we can be confident that differences found between groups are differences that have a high likelihood of being real in the American population.

Unfortunately, many media polls do not contain sophisticated measures of religion. For this reason, I only use larger, more reputable, social science surveys such as the GSS, Pew surveys and Baylor Religion Surveys. What follows, then, is a snapshot of some of the analyses that were conducted.

Circling back to the items measuring belief in science and religion conflict, I made the argument that the strongest version of this question – asking whether or not science and religion were *incompatible* – probably comes the closest to the general epistemological warfare model. Only 17% of the American public agrees that they are

**Figure 2.10: Belief that religion and science are 'incompatible' by religious tradition**

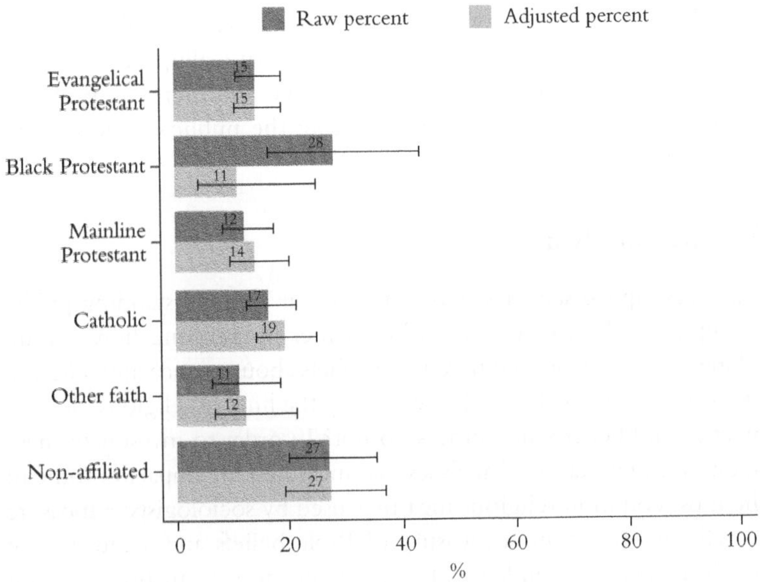


*Source:* Baylor Religion Survey II, 2007

incompatible. Does this fall along religious lines? Figures 2.10–2.12 show how this belief is different by religious tradition, religious service attendance and beliefs about the Bible. In all cases, it is those who are the least religious and the most disconnected from institutional religion and belief of any kind who are most likely to believe that science and religion are incompatible.

**Figure 2.11: Belief that religion and science are 'incompatible' by religious service attendance**

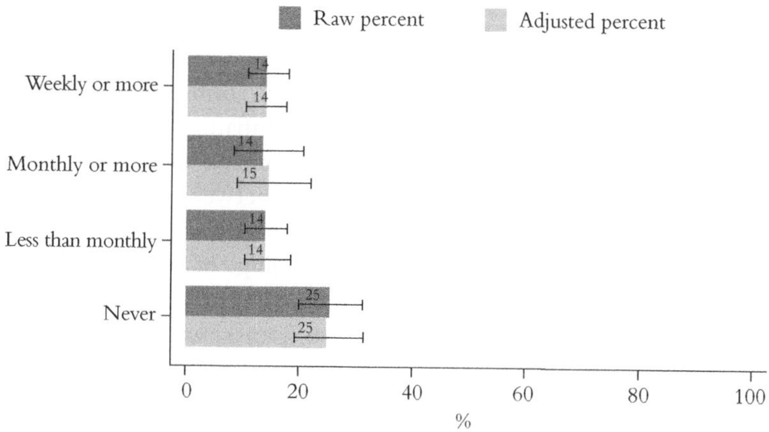


*Source:* Baylor Religion Survey II, 2007

**Figure 2.12: Percentage who believe that religion and science are 'incompatible' by Bible beliefs**

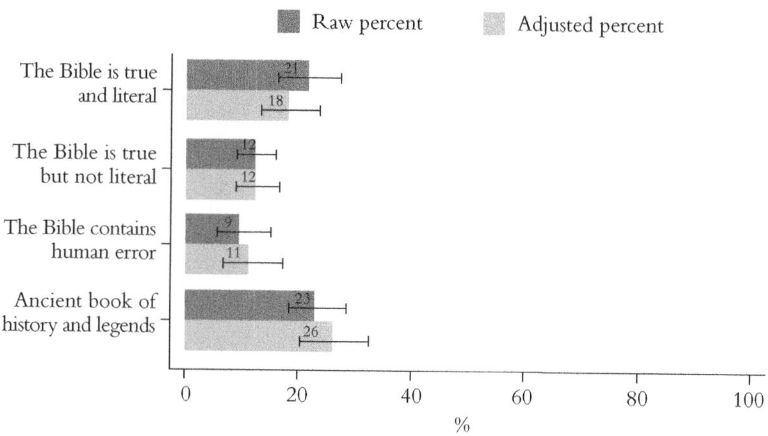


*Source:* Baylor Religion Survey II, 2007

The Bible beliefs question reveals another interesting pattern, though. Although those who believe that the Bible is a book of fables are the most likely to believe that science and religion are incompatible, the second most likely group is biblical literalists. There is a distinctive 'U-shaped' pattern to belief in incompatibility. This same pattern shows up in the 2013 wave of the NSRHO. Instead of asking whether religion and science are incompatible, this survey includes a question about compatibility. The percentage agreeing is shown in Figure 2.13, broken down by beliefs about the Bible. Again, the two groups least likely to believe that religion and science are ultimately compatible are the literalists and those who believe that the Bible has no divine influence. It should be noted that this belief about the ultimate relationship between science and religion does not tell us whether or not the American public experiences any conflict between the two. Most measures of conflict are likely conflating these ultimate ontological beliefs about the nature of religion and science with the actual day-to-day experiences of conflicts between faith and the findings of science. The following survey item better captures this latter aspect: 'Sometimes I feel like the findings of science and the teaching of [my] faith conflict with each other' (see Figure 2.14). This does *not* show the U-shaped pattern; rather, what we see is an increasing 'felt' conflict, with literalists experiencing the least conflict

**Figure 2.13: Percentage who believe that science and religion are, ultimately, compatible by Bible beliefs**

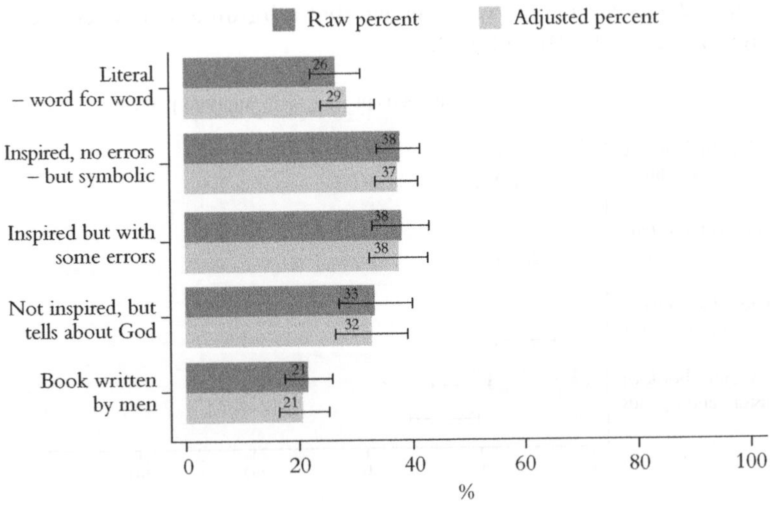


*Source:* NSRHO, 2013

**Figure 2.14: Percentage who sometimes feel like science and religion conflict by Bible beliefs**

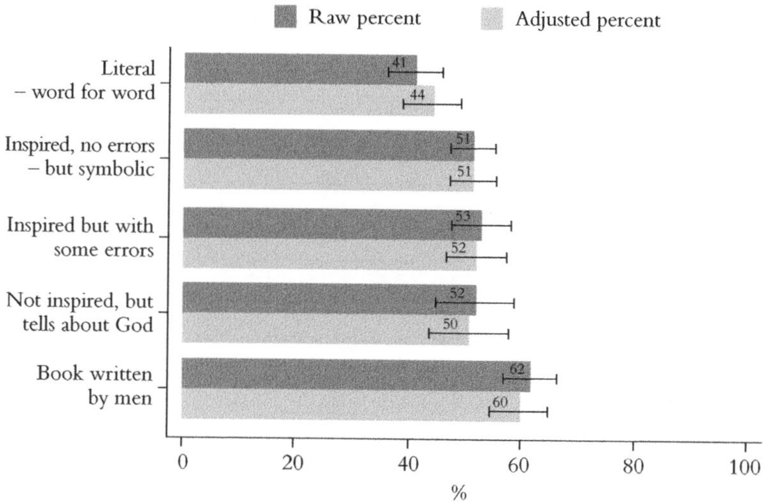


*Source:* NSRHO, 2013

(at 41%) and those at the other end of the scale experiencing the most (62%).

Moving along to human origins, we can analyse whether these same factors are important in parsing who believes in human evolution and who does not. Figures 2.15, 2.16 and 2.17 show the result of these analyses. Perhaps unsurprisingly, all of these measures are strong predictors of whether someone believes in human evolution. Although the addition of control variables weakens the direct effect (shown in the 'adjusted percent' bars), religious factors are still the primary determinants of what people believe about evolution. Race/ethnicity, income, age, gender and even political ideology all pale in comparison to the effect of religion.

There are several ways to make sense of these data. Some have claimed that religious conservatives are anti-science, that they reject rationality and empiricism for illogical leaps of faith and magical thinking (Stenger, 2012; Coyne, 2015). If this is the case, we should expect that the more conservative the Bible beliefs and religious tradition, and the more frequently one attends religious services, the more science as a whole is dismissed and avoided. However, there is little evidence that this is the case. Although the highly religious are more suspicious of scientists, science itself is rarely considered problematic. Due to space considerations, I cannot include the other

**Figure 2.15: Percentage who say that it is 'true' that humans evolved from earlier species by religious tradition**

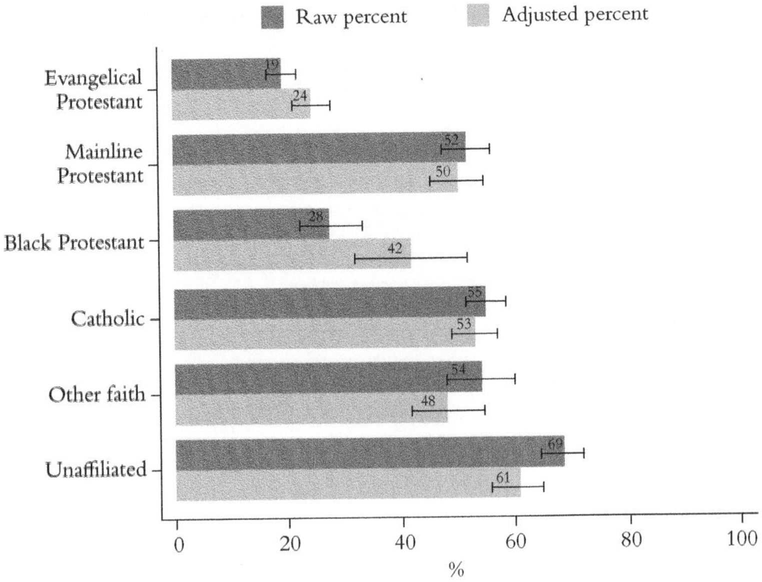


*Source:* GSS, 2006–12

**Figure 2.16: Percentage who say that it is 'true' that humans evolved from earlier species by religious service attendance**

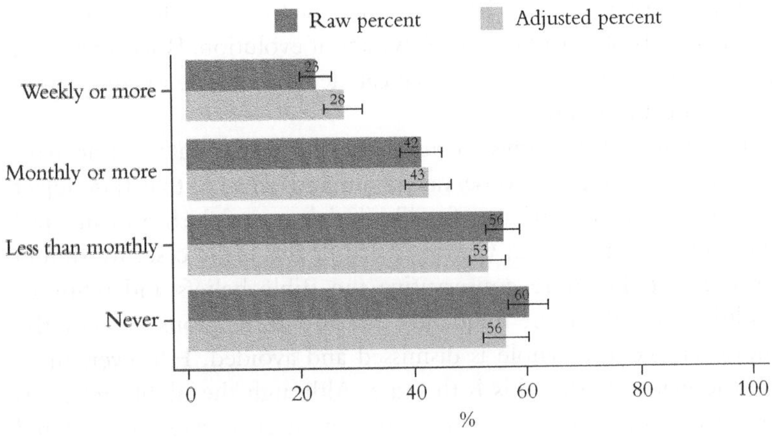


*Source:* GSS, 2006–12

**Figure 2.17: Percentage who say that it is 'true' that humans evolved from earlier species by Bible beliefs**

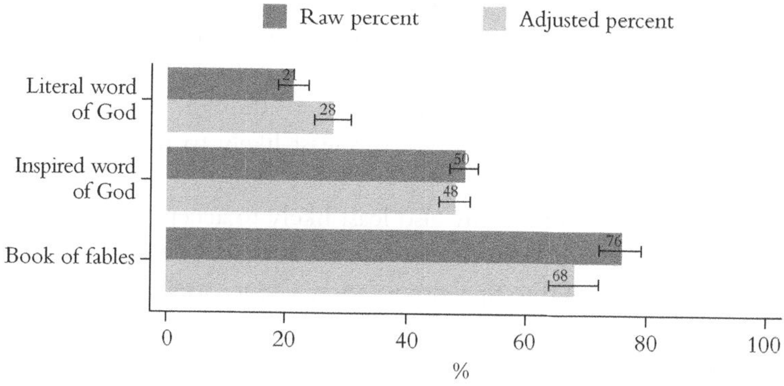


*Source:* GSS, 2006–12

measures that show no difference between the religious and the non-religious; however, I will mention a few here. Religious conservatives, for the most part, are no different from others in what they believe science consists of, who should be practising it and where this should be done (at universities, by those with advanced degrees, basing conclusions on solid evidence, examining different interpretations of results and using replication when possible). They are not very different in their hopes that science will be able to solve social problems and in their belief that scientists are working for the good of humanity. They are as likely to seek out scientific information and read about science in printed material. Clearly, religious conservatives are not simply anti-science (analyses are available from the author upon request).

There are still some additional differences, though, by religious identity, belief and practice – but most of these seem to be concerned with the ways in which scientists *practice* science and the non-scientific biases that they are perceived to have (and not science itself). For example, the belief that most scientists are hostile towards religion varies by the religion measures that we have been using. Additional data analysis finds biblical literalists, evangelical Protestants and regular church attendees are all more likely to believe that scientists are hostile towards religion (analysis not shown, but available upon request from the author).

To summarize, in this section, we were able to dig a little deeper to examine how indicators of religious belief, identity and practice help sort responses to some of the items that we have been examining. Beliefs about inherent conflict between science and religion are most likely to be adopted by those at either end of the faith spectrum (especially when measured by beliefs about the Bible). Allowing either science or religion to trump the other makes the most sense to these groups (even if it is still a minority of both groups that see science and religion as inherently incompatible). This is not the same as claiming to have actually experienced these conflicts, though. In this instance, those who are most religious are the least likely to report that they occasionally feel a conflict.

Religious conservatives are also least likely to accept evolution and are more likely to believe that some scientists harbour anti-religious biases. Despite all of this, they do not view science itself as problematic. By and large, they believe the same things as the rest of the public about what good science is, how it should be practised and the hope that it will create a better world.

## Some critiques and considerations

As this chapter has shown, quite a few surveys exist that include questions about evolution and a smaller (but not insignificant) set of surveys exist that include questions about the relationship between science and religion. Quantity is not the same thing as quality, though. Very few surveys that I have reviewed have focused exclusively on the American public's views on science, and none of the publicly available surveys reviewed have given sustained focus to evolution or the relationship between religion and science. The result is that social scientists are not presently in a good position to study many of the most interesting questions about how the public conceives of science, religion and evolution. The good news is that there are promising developments underway. In particular, the recent survey work conducted by Ecklund and colleagues (see Ecklund and Scheitle, 2018) and similar work in the UK by Elsdon-Baker and colleagues (see Elsdon-Baker et al, 2017) serve as exemplars of innovative quantitative measurement in this area. The critique and assessment that follows does not, then, view the state of affairs in the social-scientific study of religion and science as dire. Rather, what follows is my attempt to distil some general principles and suggest some directions that I think will complement and reinforce the research efforts that are already underway.

### *Description at the expense of explanation*

The goal of creating most of these measures – from the NSF-funded studies of science understanding to the Gallup questions on human origins – has been to provide descriptive tracking of the public's beliefs over time. Although I will make some methodological arguments against aspects of these measures, survey questions asked in a consistent way over time, using a probability sample of the population, allow social scientists to detect basic trends. In fact, most of the published work on the topic is just such descriptive accounts of these trends (Miller, 1998, 2004). However, perhaps the most general critique that can be levelled against these measures is their inability to provide the tools for social scientists to explore *why* certain ideas about human origins, or the belief that science and religion are in conflict, hold such sway over some segments of the population. Developing models that take into account social-psychological, social-structural and cultural factors has been difficult using the data currently available. While some movement beyond pure description has been a component of recent articles (eg Evans, 2011, 2013; Hill, 2014b; O'Brien and Noy, 2015; Noy and O'Brien, 2016, 2018; Chan, 2018), these studies invariably have to settle for the available survey measurements when far more direct measures would be ideal. Indeed, these studies frequently rely on sophisticated conceptual constructs (eg post-secularity and moral/cultural authority) but are forced to 'make do' with the measures that are presently available.

## *Replication at the expense of creativity and development*

The default for social scientists is to replicate. As a general strategy, this often makes sense. On the one hand, this practice avoids the time and effort required in developing new questions; on the other hand, it allows researchers to use what are usually considered to be valid and reliable measures outside of their area of expertise. The obvious downfall is the absence of new and interesting ways to try to capture the public's beliefs. This sort of creative activity is necessary in developing better measures. This means that there is always a trade-off between replication and advancement in measurement. Sociologists recognize, for example, that measures of religious identity used in the US were quite poor in surveys conducted in the 1970s and 1980s. They did not take seriously the historical scholarship that traced different strands of religious traditions in American history. This changed with the introduction of the RELTRAD variable to measure religious tradition in the year 2000 (Steensland et al, 2000). Now, this variable is used almost exclusively to measure religious identity in sociological research.

The primary problem, I think, is that the social-scientific study of science and religion has been almost non-existent until recently. There has not been professional energy or institutional resources expended to study this topic within the social sciences. This is quite different from the situation in other fields (namely, philosophy and history, which both have well-established research programmes for the study of science and religion). The good news is that this will likely change as social scientists are increasingly investigating this topic. Replication will certainly continue, but new and better measurements will likely be debuting as well.

### *Lack of measures on moral conceptions of science*

Most attempts to measure the public's views have been viewed in the framework of science communication. This research is primarily interested in barriers to scientific knowledge in the public, along with educational and policy strategies to increase general scientific literacy. While such work is undoubtedly important for an informed citizenry in our scientific culture, it fails to capture the broader cultural significance of public understandings of science. Ideas about science are not simply factually correct or not, but exist as part of a web of collective narratives about personhood, group belonging and social progress. While this is obvious to many in the domain of religion, it is often missed in the study of science. Yet, science, as a cultural idea, is regularly employed to do work beyond its 'textbook' definition in public life. For some, the movement of history is ultimately understood as the progressive successes and dominance of science. Ideas about DNA, neuroscience and evolution become integrated into popular cultural anthropologies. All of this suggests that for some segments of the public, certain understandings of science become ways to organize fundamental narratives about what it means to be human and what the good society looks like.

### *Mistaking elite conceptualizations of science and religion for popular conceptualizations*

Perhaps one of the reasons that many view conflicts between religion and science primarily in terms of epistemic clash and not in terms of moral conflict is because they maintain a false understanding of popular conceptions of religion and science. The sociologist John Evans has recently made this argument (Evans, 2018; see also Kaden et al, Chapter 3, this volume). It is commonly assumed that many religious conservatives are somehow anti-science because they reject evolutionary biology. In other words, the assumption is that they have some sort of coherent world view based on first principles that rejects methodological naturalism and all ideas that spring forth from such principles. However, there is little evidence that religious conservatives reject or avoid science. They are as likely to believe in, say, germ theory or the existence of atoms as anybody else. They also affirm that science is a social good and are as likely as anyone else to major in the natural sciences in college.

The reason for this is that most people are not logically consistent and coherent in their world views – religious conservatives included. This is not because the public is stupid; most simply do not have the time or incentive to develop coherent world views. Yet, the vast majority of what is 'out there' is elite dialogue about these topics. Books, web sites and blogs all represent aspects of elite discourse. Elites often assume that the public is as invested in creating coherent and intellectually defensible positions as they are, but this is simply not true. Turning to social context and group identity will be far more important in understanding why the public adopts certain beliefs and narratives about science and religion.

### *Lack of understanding of social context and group dynamics*

A related assumption, especially when it comes to accepting evolution, is that people need to be convinced that their beliefs are somehow intellectually deficient or do not correspond to the preponderance of scientific evidence. This is the reason they do not accept evolution. If this is the problem, then science education will be the primary front in any effort to change minds. However, the evidence that education actually changes what people believe is slim. Most national surveys show virtually no influence from attending and graduating from college once other factors are controlled for (Miller et al, 2006; Berkman and Plutzer, 2010). In my own analysis of who changes beliefs about evolution over time, going to college played virtually no role in this change. Creationists who attended college were actually slightly *more* likely to maintain their creationist beliefs, though the differences between them and those who never attended college were not statistically significant (Hill, 2014b). Measures of acceptance of evolution that use a pre-test and post-test format for high-school and college-level science courses show that students are frequently not persuaded (Bishop and Anderson,1990; Lawson and Worsnop,1992; Ingram and Nelson, 2006; Rice et al, 2011). This is not to say that students are never convinced – they obviously are some of the time – but the resistance to changing beliefs is strong, and the ability for people to compartmentalize their beliefs is equally strong (see the previous critique).

The evidence suggests that many make the mistake of assuming that others have an overly intellectualized conception of beliefs about religion and science, which results in an under-appreciation of the role of social context and group identity. In my analysis of who changes beliefs over time using the National Study of Youth and Religion, it was friendship networks that proved to be important, not education (Hill, 2014b). The NSRHO data show that creationists are much more likely to be in strong networks of co-religionists (both friends and family) and belong to religious congregations that have a settled anti-evolution position (Hill, 2014a). The difficulty is that there are almost no adequate measures in existing data to capture these social group dynamics. Clearly, more work needs to be done in this area.

### *Near-exclusive focus on evolution/creationism*

There is actually some good reason for this, though not enough to warrant the exclusion of other important intersections of science and religion. A 2009 Pew survey (referenced in the section on religion and science conflict) *does* indicate that when the public is thinking about conflict between science and religion, they are disproportionately thinking about the conflict between evolutionist and creationist beliefs (and by quite some margin). Nevertheless, this leaves out other arenas of potential conflict. Medicine and bioethics have received some attention in surveys (in particular, the ethics of stem cell research), but potential conflicts raised by neuroscience over ideas of mind, consciousness and free will are entirely absent from national surveys. Still other potential challenges come from evolutionary psychology on the origins of morality and altruism, as well as the science of human sexuality. Of course, one of the most notable sources of conflict is the commitment of some scientists to philosophical naturalism, which excludes belief in divine action and miracles (and has consequences for belief in the efficacy of prayer). While none of these are *necessary* sites of conflict, these are all potential areas that have mostly gone unexplored in surveys of the general public. This exclusive focus also extends to surveys in other geographical contexts (see Carlisle et al, Chapter 7, this volume), even countries where creationist organizations are much less prominent and where belief in evolution has not been a site of public conflict. Asking about a wider range of subjects, then, has potential to shed light on how the challenges and conflicts relating to science and belief vary across different societies and religious traditions.

### *Relying on questions that unintentionally prime conflict narratives*

One of the dangers of an exclusive focus on these arenas of conflict is the unintentional priming of the conflict thesis. It is very difficult to frame questions about science and religion together without framing a default position of opposition. It seems quite possible that the vast preponderance of religious experiences, ideas and beliefs, and the vast preponderance of scientific encounters and information, have almost no degree of conflict for most people. If people think back to their most recent religious worship experience, or the most recent reading of a scientific study reported in the newspaper, they are unlikely to remember encountering scientific or religious objections of one to the other. These experiences ride beneath the radar of conscious thought for most people precisely because they are so ordinary and uneventful. The few instances of perceived conflict that rise to the surface will inevitably colour any thinking about science and religion together. Public discourse and surveys that share these assumptions about conflict are likely missing everyday domains where science and religion coexist quite peaceably.

### *Relying on abstractions that are unlikely to have shared definitions among the public*

The abstraction of the concepts of religion and science in survey questions allows people to read into these terms a variety of individualized meanings and experiences. The very fact that the survey questions reviewed in this chapter were so sensitive to wording and forced-choice categories is evidence of this. The meanings that they are ascribing to these terms, and the instances of conflict or consensus that are brought to mind, will vary from person to person.

One potential solution is to increase vignettes in survey questions. When the meanings imputed to ideas – such as religion and science – vary substantially, short vignettes that describe hypothetical behaviour and beliefs, and ask the respondent to assess or react to the story, will often be more telling than an abstract, contextless question. There is a growing literature on using 'anchoring vignettes' to help ground survey responses in areas such as health, political efficacy and job satisfaction (King et al, 2004; King and Wand, 2007; Hopkins and King, 2010). Perhaps extending this method to science and religion research would be a productive way to get 'underneath' the ideas that the public draws upon when thinking about religion and science.

The second suggestion is to disaggregate complex beliefs into their component parts. I did something along these lines in the NSRHO (Hill, 2014a). Instead of asking questions designed to categorize respondents as creationists, theistic evolutionists or non-theistic evolutionists (as the frequently cited Gallup poll question does), I asked a series of questions about evolution, God's involvement in the emergence of humans, the historical existence of Adam and Eve, 24-hour days of creation, and geological time frame. These simpler questions become building blocks to more accurately construct complex beliefs and identities. Using this method, I found that very few people fit unambiguously into camps such as 'Young Earth creationists' or 'non-theistic evolutionists'.

## Conclusion

This chapter reviewed the available data in US national surveys on science and religion conflict, human origins, and beliefs about scientists. Although there are a number of measures that are probably adequate for basic descriptive tracking of the beliefs of the public over time, sociologists are not able to use most of these data to understand some of the most pressing questions: why does the idea of an inevitable warfare between science and religion continue to hold such sway over segments of the population when historians have largely debunked it? What social factors are key to understanding when and why some people change their beliefs about human origins? What role does scientism play in perpetuating the perception of opposition between religion and science? What role do religious communities play in fomenting distrust between religious conservatives and scientists?

The solution is to be more intentional about the data collected. Each of the critiques outlined in the previous section contains either explicit or implicit fixes. Measures need to be more carefully calibrated to how the public actually comes to think about these issues. New questions about morality, progress and teleology (for *both* religion and science) need to be generated. Measures of social context (friends, family, congregations), and group identity and dynamics, should be included. Careful attention to domains of conflict outside of human origins, and creative attempts to avoid unintentionally priming conflict between religion and science, should be a part of any data-collection enterprise. The good news is that much of this scholarship is already underway. Large-scale survey analyses by Ecklund and colleagues (see Ecklund and Scheitle, 2018) and Elsdon-Baker and colleagues (see Elsdon-Baker et al, 2017) are providing new models for the quantitative measurement of public attitudes towards science and religion. Furthermore, many of the contributions to this volume directly address the shortcomings that I have outlined. If this work is any indication, the coming years should be a particularly generative time for the social-scientific study of science and religion.

## Notes

1 For an overview of some of these survey items, see: http://thearda.com/MAWizard/Concepts/MW_34_CC.asp

2 Some versions of this question, such as ones in the Baylor Religion Survey and Pew Religious Landscape survey, ask to what extent respondents agree or disagree with the statement about human evolution. Again, four answers are used that range from 'strongly' or 'completely' agree to 'strongly' or 'completely' disagree. The results are similar to the questions that ask whether the statement is true or false.

## References

Baker, J.O. (2012) 'Public perceptions of incompatibility between science and religion', *Public Understanding of Science*, 21(3): 340–53.

Bauer, M.W., Allum, N. and Miller, S. (2007) 'What can we learn from 25 years of PUS survey research? Liberating and expanding the agenda', *Public Understanding of Science*, 16(1): 79–95.

Berkman, M. and Plutzer, E. (2010) *Evolution, creationism, and the battle to control America's classrooms*, New York, NY: Cambridge University Press.

Bishop, B.A. and Anderson, C.W. (1990) 'Student conceptions of natural selection and its role in evolution', *Journal of Research in Science Teaching*, 27(5): 415–27.

Chan, E. (2018) 'Are the religious suspicious of science? Investigating religiosity, religious context, and orientations towards science', *Public Understanding of Science*, 27(8): 967–84.

Coyne, J.A. (2015) *Faith versus fact: Why science and religion are incompatible*, New York, NY: Penguin.

Davis, R.C. (1957) *The public impact of science in the mass media*, Ann Arbor, MI: University of Michigan Institute for Social Research.

Dickson, D. (2005) 'The case for a "deficit model" of science communication'. Available at: www.scidev.net/global/communication/editorials/the-case-for-a-deficit-model-of-science-communic.html (accessed 3 October 2018).

Ecklund, E.H. and Scheitle, C.P. (2018) *Religion vs. science: What religious people really think*, New York, NY: Oxford University Press.

Elsdon-Baker, F., Leicht, C., Mason-Wilkes, W., Preece, E. and Piggot, L. (2017) *Science and religion: Exploring the spectrum: Summary report of preliminary findings for a survey of public perspectives on evolution and the relationship between evolutionary science and religion*, Birmingham: Newman University and YouGov. Available at: https://sciencereligionspectrum.org/wp-content/uploads/2017/09/SRESYouGov-survey-preliminary-findings-5.9.17.pdf (accessed 27 September 2018).

Evans, J.H. (2011) 'Epistemological and moral conflict between religion and science', *Journal for the Scientific Study of Religion*, 50(4): 707–27.

Evans, J.H. (2013) 'The growing social and moral conflict between conservative Protestantism and science', *Journal for the Scientific Study of Religion*, 52(2): 368–85.

Evans, J.H. (2018) *Morals not knowledge: Recasting the contemporary US conflict between religion and science*, Berkeley, CA: University of California Press.

Gould, S.J. (1997) 'Nonoverlapping magisteria', *Natural History*, 106(2): 16–22.

Higgins, W.B., Thompson, R.K. and Bishop, G. (2011) 'Evolving a better measure: developing alternative measures of beliefs in human origins', paper presented at the 66th Annual Conference of AAPOR, Phoenix, AZ, 12–15 May.

Hill, J.P. (2014a) *National study of religion and human origins*, Grand Rapids, MI: BioLogos.

Hill, J.P. (2014b) 'Rejecting evolution: the role of religion, education, and social networks', *Journal for the Scientific Study of Religion*, 53(3): 575–94.

Hopkins, D.J. and King, G. (2010) 'Improving anchoring vignettes: designing surveys to correct interpersonal incomparability', *Public Opinion Quarterly*, 74(2): 201–22.

Ingram, E.L. and Nelson, C.E. (2006) 'Relationship between achievement and students' acceptance of evolution or creation in an upper-level evolution course', *Journal of Research in Science Teaching*, 43(1): 7–24.

King, G. and Wand, J. (2007) 'Comparing incomparable survey responses: evaluating and selecting anchoring vignettes', *Political Analysis*, 15: 46–66.

King, G., Murray, C.J., Salomon, J.A. and Tandon, A. (2004) 'Enhancing the validity and cross-cultural comparability of measurement in survey research', *American Political Science Review*, 98(1): 191–207.

Lawson, A.E. and Worsnop, W.A. (1992) 'Learning about evolution and rejecting a belief in special creation: effects of reflective reasoning skill, prior knowledge, prior belief and religious commitment', *Journal of Research in Science Teaching*, 29(2): 143–66.

Masci, D. (2009) 'Public opinion on religion and science in the United States', Pew Forum on Religion and Public Life, 8(15). Available at: www.pewforum.org/Science-and-Bioethics/Public-Opinion-on-Religion-and-Science-in-the-United-States.aspx

Miller, J.D. (1998) 'The measurement of civic scientific literacy', *Public Understanding of Science*, 7: 203–23.

Miller, J.D. (2004) 'Public understanding of, and attitudes toward, scientific research: what we know and what we need to know', *Public Understanding of Science*, 13(3): 273–94.

Miller, J.D., Prewitt, K. and Pearson, R. (1980) *The attitudes of the US public toward science and technology: Analytic report*, Chicago, IL: National Opinion Research Center, University of Chicago.

Miller, J.D., Scott, E.C. and Okamoto, S. (2006) 'Public acceptance of evolution', *Science*, 313(5788): 765–6.

Noy, S. and O'Brien, T.L. (2016) 'A nation divided: science, religion, and public opinion in the United States', *Socius*, 2, DOI:10.1177/2378023116651876.

Noy, S. and O'Brien, T.L. (2018) 'An intersectional analysis of perspectives on science and religion in the United States', *The Sociological Quarterly*, 59(1): 40–61.

O'Brien, T.L. and Noy, S. (2015) 'Traditional, modern, and post-secular perspectives on science and religion in the United States', *American Sociological Review*, 80(1): 92–115.

Rice, J.W., Olson, J.K. and Colbert, J.T. (2011) 'University evolution education: the effect of evolution instruction on biology majors' content knowledge, attitude toward evolution, and theistic position', *Evolution: Education and Outreach*, 4(1): 137–44.

Scheitle, C.P. (2011) 'US college students' perception of religion and science: conflict, collaboration, or independence? A research note', *Journal for the Scientific Study of Religion*, 50(1): 175–86.

Scheitle, C.P., Johnson, D.R. and Ecklund, E.H. (2017) 'Scientists and religious leaders compete for cultural authority of science', *Public Understanding of Science*, 27(1): 59–75.

Steensland, B., Robinson, L.D., Wilcox, W.B., Park, J.Z., Regnerus, M.D. and Woodberry, R.D. (2000) 'The measure of American religion: toward improving the state of the art', *Social Forces*, 79(1): 291–318.

Stenger, V.J. (2012) *God and the folly of faith: The incompatibility of science and religion*, Amherst: Prometheus Books.

3

# Language, Labels and Lived Identity in Debates about Science, Religion and Belief

*Tom Kaden, Stephen H. Jones and Rebecca Catto*

## Labels and their functions

Public discourse[1] about science and belief is permeated by all manner of labels: terms like 'creationism', 'intelligent design', 'Darwinism' and 'New Atheism'.[2] Some of these labels describe a belief about evolution. Others signify a conviction about how science and religion relate. Still others describe an organization, social movement, cultural trend or group of people. In a few cases, the same label (eg 'creationist' or 'New Atheist') serves all of these functions, with the term being used to describe both a set of beliefs and the population that supposedly holds these beliefs. Labels also, as we will see, feature in social-scientific research, where they often form the basis of survey questions designed to measure people's understanding and acceptance of aspects of science. What is not typically questioned, however, is what people actually *think* about such labels. Are people aware of these terms? Do they identify with them, referring to themselves as 'creationists' or 'New Atheists'? Do these labels accurately represent people's perspectives? Often, in both academic and popular discussions of science and belief, the implied answer to questions such as these is 'Yes'. Commentators commonly describe certain religious groups as dominated by 'creationists', in the process assuming that this population holds a certain set of beliefs. This chapter, however, challenges such assumptions. Building on an emerging body of sociological scholarship (Evans, 2011, 2018; Hill, 2014; Kaden et al, 2017; see also Hill, Chapter 2, this volume), we highlight a gap between scholarly and popular categorizations of evolution belief and the views that people actually hold.

We look at the way in which labels were used in 123 interviews and 16 focus groups carried out with members of the public and scientists working in the life, biological and medical sciences in the UK and Canada. In these interviews, which focused on the theme of religion and evolution, we asked participants whether they were aware of, or identified with, any labels used to describe belief about evolution, and examined the way participants did, or did not, use terms such as 'creationism', 'intelligent design', 'Darwinism' and so on. Using these data, the chapter advances three related arguments. The first is that, in general, people who are not involved with organizations directly concerned with science and religion-related questions (and even some of those who are) do not identify with commonly used labels. Our research participants recognized such labels only selectively and affiliated with them even less. This can be said of both members of the general public and professional scientists, though recognition of technical terms was certainly higher among the latter group. The chapter thus expands and gives further backing to US-based studies (Hill, 2014; see also Hill, Chapter 2, this volume) which have shown that the arguments of creation science organizations are not as influential as commonly thought.

The second argument is that not only do people rarely self-identify with labels, but there are also profound dangers in imposing such labels upon groups of people as etic descriptive categories. Many science and religion-related labels, while often treated as neutral terms by journalists and researchers, have been developed and used strategically in public life to advance a political agenda, predominantly but not exclusively in the US. Using them carelessly to categorize groups of, say, evangelical Christians or non-religious people can exaggerate the influence and significance of professional anti-evolution or secularist organizations by implying that they have a stable cohort of 'followers' (Elsdon-Baker, 2015). We do not argue that terms like 'creationist' or 'Darwinist' should never be used, though we argue that some refinement and creative thinking is needed – dispensing with such terms altogether would likely sow even more confusion. Rather, we argue that social scientists should work to generate more refined descriptions of groups and be attentive to *how* labels concerning science and belief are used.

This last point leads on to our third argument, namely, that labels are *creatively engaged with*. Our participants' responses to labels were not just a matter of accepting or rejecting them. People redefined labels for themselves, objected to some of their public uses or rejected some of the knowledge claims that are commonly associated with them. In our research, while people did not typically 'fit' or identify with science and belief labels, some participants did use them to position themselves: labels sometimes functioned as discursive resources in the construction, remaking and articulation of personal identity, and this includes contesting common assumptions about what (religious and secular) communities hold to be true. Social scientists, we therefore argue, should be alert to the language games that are played not only in public spheres by organizations seeking to win public recognition for ideas, but also in people's everyday lives. Remaining aware of these tendencies, we conclude, is needed if social scientists are to avoid unintentionally reinforcing the work of interest groups active in science and belief debates.

## Public development and contestation of evolution labels

Public debates about science and religion, especially those focusing on human origins, are often configured around the terms *creationism* and *evolution*. Both of these terms can be subcategorized in numerous ways, some of which complicate the binary opposition. Among the forms of creationism are a number of labels that describe different forms of Biblical hermeneutics and the connections drawn from Biblical interpretation to various fields of modern science. 'Young Earth creationism' is based on Biblical inerrantism (Numbers, 2006: 73, 85, 251), which leads its proponents to read much of the text literally, and to conclude that the universe is just 6,000–10,000 years old. This, in turn, leads them to question vast areas of established scientific knowledge in the fields of biology, astronomy, geology, anthropology and physics, especially the idea of evolution by natural selection. This position contrasts with 'Old Earth creationism', which contains a range of different forms of non-literalist Biblical interpretation that grant more time for natural processes in the framework of Biblical creation. Among these forms is day/age creationism, which posits that the days of creation can be understood as epochs of time, and progressive creationism (Ramm, 1981 [1954]), which holds that God created different species consecutively throughout a long history of the earth. Then, surrounding the core label of creationism are positions that are regularly invoked alongside creationist positions but for which the accuracy of the label is questionable. 'Geocentrism' and 'flat-eartherism', for example, are two marginal positions that tend to be used for polemical purposes in public debate that go beyond what is generally associated with creationism. Often aligned, too, is intelligent design (Behe, 2006; Meyer, 2010), which holds that certain biological structures are best explained by a designing intelligence. This position is, however, distinguishable by the fact that it is removed from the central creationist principles of Biblical hermeneutics.

Evolution, the second core label in public debates surrounding science and religion, contains the same diversity. Sub-labels that fall under this heading are generally distinguished not by beliefs about evolution – on which there is broad agreement in the public discourse[3] – but by the way in which they relate to religion. Best known among the positions that have a negative relationship with religion are 'scientific atheism' or 'New Atheism', both of which hold that a scientific outlook on the world entails not only the rejection of creationism, but of religion altogether. The term New Atheism was coined to describe a group of authors and activists many of whom are not primarily associated with science advocacy (Kettell, 2013). Nevertheless, evolution features as a key concept in this position since, in the words of one of the best-known New Atheists, Richard Dawkins (1986: 6), 'Darwin made it possible to be an intellectually fulfilled atheist'. These arguments contrast with positions that seek 'neutrality' with respect to religion, notably, 'NOMA' (or the principle of 'non-overlapping magisteria'), which holds that science and religion will not conflict as long as it is recognized that they address different questions. The term 'NOMA' was coined by the palaeontologist Stephen Jay Gould (2002), and its current defenders include organizations such as the National Center for Science Education (NCSE), whose main focus lies on evolution education in the US. These two subgroups of advocates for evolution are then joined by a third, which consists of positions that include some form of religious outlook in an evolutionary framework. The most prominent among those labels are 'theistic evolution' and 'evolutionary creationism', the official position of the BioLogos Foundation, a non-profit Christian organization founded by Francis Collins, the American physician and geneticist who led the Human Genome Project and is presently director of the National Institutes of Health in the US.

The functions that labels maintain in public discourse are manifold. In some cases, they are used by groups and individuals to self-identify. For instance, Reasons to Believe, an advocacy organization for Old Earth creationism, uses this label to describe its own position.[4] In other cases, parts of labels are accepted but others rejected. For instance, the largest creationist organization in the US, Answers in Genesis, labels itself creationist, but even though it advocates a Young Earth position, it is critical of the label 'Young Earth creationism'. Instead, its founder and chief executive officer (CEO), Ken Ham, labels his own position as 'revelationist, no-death-before-Adam redemptionism' (quoted in Answers in Genesis, no date[a]). Even within the core institutions that represent labels, then, there are tendencies to criticize and redefine labels, which we shall encounter later. This extends to labels used to identify public advocates of atheistic evolution as well. For instance, Richard Dawkins, who is regarded as the prime spokesperson for New Atheism in the public sphere, actually rejects this term entirely since he does not see any major differences between his views and earlier forms of atheism (Dawkins, 2016: 5; for a full sociological analysis of the distinctions between 'old' and 'new' atheism, see Kettell, 2013).

From these examples, we can conclude that the function of labels as self-descriptors is, at best, ambiguous. To an extent, this ambiguity relates to how evolution labels are also used to characterize *others* in pejorative terms. In some cases, the ascription of labels to movements or people is balanced. Answers in Genesis, for example, delineates the differences between its Young Earth position and intelligent design in simple descriptive terms (Answers in Genesis, no date[b]). Frequently, however, the intention behind labelling others is to delegitimize them. Here, too, intelligent design is a good example since one of the main strategies used to counter its public influence has been to label it as 'creationism in a cheap tuxedo' (Melott, 2002; see also Pennock, 2001). The 1987 US Supreme Court case Edwards vs Aguillard ruled that inserting scientific creationism into school curricula violated the First Amendment's 'Establishment Clause'. These criticisms, then, can be seen as 'boundary work' (Gieryn, 1999), whose aim is to ensure that intelligent design is also consigned to the spheres of church and family (Scott and Branch, 2006).[5] The same strategic use of labels can also be observed in the 'other direction': representatives of intelligent design commonly label New Atheists and other evolutionary biologists as 'Neo-Darwinists', the intention being to characterize these groups and individuals as representatives of merely one competing scientific paradigm and intelligent design as a legitimate, though marginalized, contender.

Another broad function of labels within the public discourse is to establish a relative order of positions. Different variants of creationism are sorted and compared with regard to the extent to which they are at odds with current science, or depending on the hermeneutical principles that they apply. One well-known example of such ordering is Eugenie C. Scott's (2009) 'creation–evolution continuum', in which the former director of the National Center for Science Education ranks positions from 'flat earthers' through 'Young Earth creationism' and 'theistic evolution' to the implied endpoint, 'atheistic evolution' (see Figure 3.1). Finally, labels can simply be applied not only to organizations representing particular stances, but also to particular groups of people. Dawkins, for example, has said that in the UK, 'most devout Muslims are creationists' (Gardham, 2008). Leaving to one side the lack of evidence available to back up this claim (Carlisle et al, Chapter 7, this volume), what is notable here is how the label is used to describe Muslims rather than the religious tradition – Christianity – out of which creationism emerged. Similarly, in Canadian public debates, 'evolutionism' has become part of a generalized anti-religious rhetoric for people identifying with the New Atheist movement (LeDrew, 2016).

**Figure 3.1: The creation–evolution continuum, by Eugenie C. Scott**

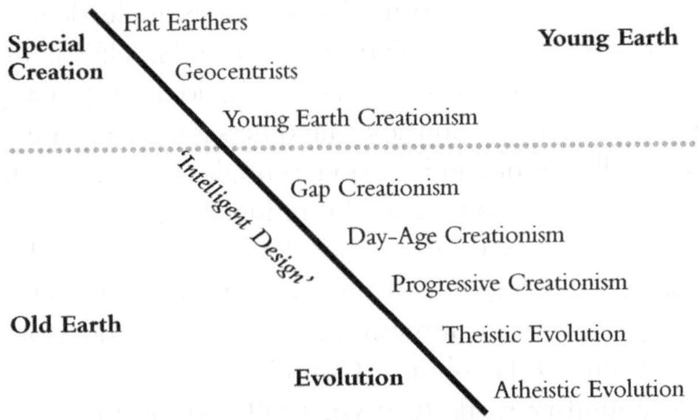


## Labels in social-scientific research

With the language of evolution belief being so contested, offering a 'neutral' description of organized movements, and even more so people's beliefs, becomes highly challenging. Labels are used to delineate what researchers study and, voluntarily or not, these descriptions also entail statements of relative order (Aechtner, 2016; Numbers, 2006). For instance, the second edition of Numbers's (2006) groundbreaking historical study of creationism contains a section on intelligent design. Numbers makes clear that intelligent design is part of the history of creationism, even though representatives of intelligent design, as well as many self-described creationists, would object to this. As with the case of Scott mentioned earlier, the use of labels entails an implicit or explicit analysis and evaluation of the terms applied, even if this is not the intention.

These difficulties are perhaps most pronounced in relation to quantitative surveys. Labels are sometimes used by social scientists as items or parts of items in surveys designed to measure the public's acceptance of positions on science and religion. For instance, in a 2006 UK poll commissioned by the BBC, Ipsos MORI asked specifically which of the following statements best described respondents' views (see Elsdon-Baker, 2015: 26):

- The 'evolution theory' says that humankind has developed over millions of years from less advanced forms of life. God had no part in this process.
- The 'creationism theory' says that God created humankind pretty much in his/her present form at one time within the last 10,000 years.
- The 'intelligent design' theory says that certain features of living things are best explained by the intervention of a supernatural being, for example, God.

The influence of labels in surveys is generally not as clear. Even when they are not explicitly mentioned, however, labels are often used as templates for the options presented to respondents, with answers corresponding to systematic positions argued by Young Earth creationists or theistic evolutionists (see Hill, Chapter 2, this volume). The problem with this is that the available evidence from more sophisticated quantitative research clearly demonstrates that the relationship between labels and the sets of beliefs they supposedly signify is limited. Many self-labelled 'creationists' do not assent to all, or even most, truth claims put forward by the public figures and groups generally seen as representing creationism (Hill, 2014). There are also more limited indications that publics are, consciously or unconsciously, aware of the pejorative and political connotations of labels associated with evolution belief (see Hill, Chapter 2, this volume). Referring to 'Darwinism' in surveys conducted with religious people, for example, appears to lower rates of evolution acceptance (Kuebler, 2016). Using positions that stem from public discourse and simply polling consent to these, then, runs the risk of, in Elsdon-Baker's (2015) words, 'creating creationists'.

Beyond this issue of misrepresentation is the problem of not engaging with what actually concerns publics when they assent to or reject labels or positions on science and religion. 'Creationism', 'intelligent design', 'New Atheism' and so on are, first and foremost, signifiers of sets of propositional beliefs, and their creators put emphasis on this aspect since they are interested in presenting their views as true. However, ethnographers (Toumey, 1994; Long, 2011) and sociologists (Evans, 2011, 2018) have shown that among professionals and publics, there is a moral dimension that motivates the development and uptake of propositional stances, and consequently assent to labels. By focusing on labels, or the key propositional aspects of positions that are signified by them, public polling might miss what people are actually referring to when they agree or disagree. There is incipient research showing that this gap between professional argumentation and public opinion actually exists. In Michael S. Evans (2016) recent mixed-methods study, when members of the American public were asked whether people feel represented by public figures – that is, those who are most closely identified with public labels – they overwhelmingly answered in the negative. Although not directly concerned with labels, Evans's study confirms that the focus on propositional stances leaves unattended a range of non-propositional factors that might influence public identification with labels.

Misreading popular identification with labels and the propositions associated with them can have public consequences. The use of labels in popular media and research can easily lead to the overstatement of public knowledge concerning different positions on the relationship between science, religion and evolution. A striking example of this is the controversy surrounding the adoption of intelligent design in the curriculum of public schools in Ohio in 2002. Polls at that time indicated majority support for the teaching of intelligent design. George Bishop (2003), however, a political scientist suspicious of the use of labels in these polls, asked respondents whether they knew anything about intelligent design; 84% of respondents said they did not. Beyond this basic issue are problems of coherence. Most labels refer to systems of beliefs that are mutually exclusive. In the mind of their creators, one cannot be Young Earth creationist and Old Earth creationist at the same time, or support intelligent design and theistic evolution simultaneously. This might not be true for members of the general public due to their selective engagement with the epistemic content of these labels. Lastly, using labels to create a template while measuring the popularity of views on science and religion, as well as to assess the state of the public debate, might inhibit the quality of this debate. Such an application of labels puts the emphasis on doctrinal differences and obscures moral differences, commonalities and fluidity.

## Research methods

In this chapter, we do not want to argue against the use of labels in social science. Rather, we argue that the issues described earlier invite further study and refinement of methods and categories when researching science and belief. By examining how publics and professional scientists make use of labels, we can learn something not only about identification and dis-identification with propositional beliefs and cultural movements, but also about how individuals relate to the terms and categories in terms of which (Western, anglophone) public discourse is conducted. At the very least, fleshing out the complexity of the use of labels tells us why, and in what way, social scientists need to be cautious when using labels as signifiers of the public's positions on science and religion.

Our data come from the large-scale, multinational and multidisciplinary research project 'Science and Religion: Exploring the Spectrum' (2014–17). In the course of this project, we conducted 123 semi-structured interviews and 16 focus groups with members of the public and life scientists from the UK and Canada. The project aimed to investigate how debates about and perceptions of evolution, as well as science and religion more generally, are socially embedded and influenced by social context and public narratives. Our data set is split approximately 1:1 between the UK and Canada[6] and 2:1 between members of the general public and professional scientists. The sample consisted of approximately the same number of religious and non-religious participants.[7] We selected participants purposively to ensure heterogeneity in terms of religion, age, class, gender and ethnicity (Emmel, 2013). In the case of interviews and focus groups with members of the public, this was facilitated by the use of a pre-screening survey that allowed participants to be selected on the basis of the characteristics described, as well as their interest in science and/or religion. The pre-screening survey was employed to avoid recruiting too many individuals who maintain a personal agenda in public debates on science and religion. All respondents had completed education at least up to age 18 and claimed some awareness of evolutionary theory. This purposive approach, and an over-representation of educated individuals in our sample, do limit the scope of our findings somewhat. Yet, given the range of the sample, we are confident that we are not overstating the salience of labels.

Interviews covered participants' (non-)religious backgrounds, interest in the sciences, views on the subject of science and religion, and perceptions of public coverage of science and religion. Interviews were semi-structured, following a standard schedule that enabled participants to elaborate their views and draw associations between themes (Wengraf, 2009). As noted earlier, the interview schedule included a direct question asking participants whether they were aware of, and identified with, any particular labels describing positions on evolution. This was asked as an open question to begin with, which was then followed with prompts ('Darwinist', 'creationist', 'evolutionist', 'theistic evolutionist') if the interviewee was not able to name any labels.

We recorded and transcribed the interviews and focus groups. The transcripts were then uploaded to and analysed using the qualitative data analysis program NVivo, with coding performed by the interviewers (Kaden and Jones). We initially coded a sample of transcripts inductively and then agreed the coding scheme together, sustaining continuous dialogue between project team members in order to ensure consistency between coders (Campbell et al, 2013). This chapter is based on analysis of a tree node on 'language and labels', as well as nodes relating to specific terms ('creationist', 'New Atheism', etc), that emerged through coding and analysing the data. In what follows, we use this analysis to ask: how do scientists and publics make use of popular labels that delineate positions on science and religion?

## Findings

### *Knowledge and relevance of labels*

Our first question is: what prior knowledge of labels did participants have? Overall, awareness of labels in our sample was limited. A notable minority of our interviewees had no, or only very vague, knowledge of labels and their content. There was greater awareness among our scientist interviewees, perhaps unsurprisingly, as well as Canadian interviewees (though it is hard to confirm if this contrast was related to national context or the nature of our samples). Lack of knowledge extended to a clear majority of participants in the case of technical terms and subcategories such as 'theistic evolution' or 'Old Earth creationism', but it included a notable minority even in the case of common terms such as 'creationism', 'intelligent design' or 'New Atheism':

> INTERVIEWER: "Have you ever heard of intelligent design?"
> RESPONDENT: "I'm not sure. Is that a religious concept?" (Lindsay,[8] insurance broker, Christian, Canada)
> INTERVIEWER: "Do you ever find yourself kind of aligning with, or in agreement with, those people in public who go under the heading of 'New Atheists', who sort of …
> RESPONDENT: "I don't really, I have never heard of that term." (Jessie, post-doctoral psychologist, atheist, UK)

In our interviews, the clearest example of a lack of awareness of science and religion labels was Julie, a Belgian-born UK resident and follower of Christian science who was studying sociology as a mature student at the time of interview. Julie asked to see our question schedule prior to her interview.[9] Reading this list initially caused Julie to withdraw from the interview on the basis that she was not aware of terms like 'creationism' (listed as one of our conversation prompts) and thus did not feel 'expert' enough to answer questions on this theme (see also Reid, Chapter 4, this volume).

In addition to these examples of an apparent lack of awareness of, and interest in, labels, we also encountered more nuanced cases. Some of our respondents expressed familiarity with some labels while also admitting to being largely unacquainted with others, even though these labels and their proponents are in constant dialogue in public discourse. In most cases, this was dependent upon various factors associated with the biographies of interviewees, including the religious traditions in which participants were raised, as well as their professions. Arabella, for instance, is a community worker from Canada and belongs to a conservative Christian church that teaches Young Earth creationism as one of its core tenets. She was therefore very much aware of that term, along with intelligent design, although her assessment of intelligent design is guided by the supportive links she sees between it and Young Earth creationism:

> "A lot of the arguments or writings from intelligent design a lot of times I can say, like, 'Yeah, I agree with that'.… It's interesting that it can … you know, similar ideas as Young Earth creationism can come from people who aren't necessarily coming at it from, you know, 'The Bible says it, therefore …'. So, I didn't really get into the idea of intelligent design but it seemed to confirm, or be compatible with, the Young Earth creationist viewpoint,

> but … that's not really a big issue in my circles. I don't know … anyone who holds that view specifically."

In Arabella's case, the relevance of the label of intelligent design for her was dependent on its relevance to Young Earth creationism.

Biographical factors also played a central role in many cases where interviewees had high levels of knowledge of evolution labels. For example, Max, a retired Canadian pastor who subscribes to theistic evolution, demonstrated familiarity with intelligent design in his interview, though he was unconvinced of its merits:

> "I call it 'candyfloss': you bite into it and there's not much there. But, you know, it's tried … I tried to have a … what's called a 'big tent', bring together Christians from various ideas and bring them together under one common view. But it's very confusing because there's people like Michael Behe and this irreducible complexity, which to me is a 'God of the gaps' argument that says wherever science can't explain something, that must be where God is."

Like Arabella, Max came to engage with intelligent design as a result of a specific religious interest. Max was attracted by the idea that intelligent design might serve as a 'big tent' that unifies different strands of Christianity. Yet, where Arabella saw elements in intelligent design that support her position, Max came to reject it, mainly because he perceived it to be at odds with scientific consensus. The very elements that make Arabella supportive of intelligent design repel Max from it.

Throughout his interview, Max displayed a notable awareness and in-depth knowledge of all major labels and positions concerning religion and evolution that are used in public discourse, making him exceptional among our interviewees. Max's rather uncommon position results from a personal crisis. Having served as a pastor in a conservative Canadian church, he came to abandon Young Earth creationism several years previously, though it remains part of the church's doctrine. Despite the change in his own views, he continued in his vocation at the church as a covert 'heretic'. Throughout these years, Max used the Internet to explore positions on science and religion in order to find a stance more acceptable to him than the creationism he had come to reject. He also took up correspondence with many scientists and other people engaged in public debates about science and religion. This virtual community, and his intense engagement with labels, served as a counterweight to the views expressed by his church's congregation.

In Max's case, and in the small number of similar examples across our sample, the perceived need to engage with labels seems to have arisen as a consequence of personal challenges, in this case, the need to utilize resources in remaking his world view.

### *Acceptance and rejection of labels*

We might then ask: to the extent that participants were conscious of labels, did they identify with or otherwise consider them to be useful? When discussing labels, many of our respondents presented general views on the usefulness of labels and why they accept them or not. Some, albeit a small minority, expressed a positive attitude. This was the case for Helen, a Canadian Christian and teacher at a Catholic high school:

> "I think it's good if you can narrow down your views and be more specific about exactly what you believe, because I don't think we should just wipe out those labels because I think they help us understand what we believe. I don't think it would be possible to understand exactly what we believe unless we have some sort of narrowing down. But people are coming up with new ideas as we speak, so there will probably be more labels. I think sometimes labels can divide people and make people go against each other; but it does also kind of categorize things and help find patterns. So, I don't know; there's pros and cons."

While Helen displayed a broadly positive attitude, the clear majority of those who had an opinion were more critical about the existence and function of labels, and they mentioned a variety of reasons for this. For some participants, the point Helen makes – namely, that labels divide people and do not do justice to individual stances – is key to their rejection. This is the case for Lucas, an engineer living in the UK, who observed:

> "I think everybody comes to [science and religion] with a very personal experience.... I don't think labelling people is all that helpful really. I mean we have to expect that we'll have a subtle range of opinions which need to be honoured, basically."

Other respondents were reluctant to use labels for the similar, but more personal, reason that they do not feel that they fit into the categories available. Katie, for example, is a psychology student from the UK. She was raised in a household she describes as 'fundamentalist Christian' but drifted away from this in early adulthood, though she still identified in interview as an evangelical Christian. While going through this change and working out the implications for her relationship with peers and family, she, like Max mentioned earlier, acquired an awareness of labels. However, while she professes a belief in 'creation', her views are not specific enough to map onto a single label:

> "No, I think [science and religion are] intricately linked so because I have a belief that there is a god … so I believe in creation. I don't know how, but I think that God overruled it in whatever way that he chose to. I don't know that I'd say I was a six-day creationist but I also wouldn't call myself a theistic evolutionist. So, my understanding is that everything that we see and we can know about this world tells us something about who God is."

Here, Katie finds herself unable to identify with *any* label. A more common approach was for interviewees to take issue with one specific label. In some cases, this opposition to a label was intellectual, with interviewees rejecting a specific moniker because they doubted the validity of the position that it signifies. For example, Douglas is a Canadian computer programmer and self-described secular humanist. He was critical of the label of NOMA because he sees it as factually wrong:

> "I don't find that science ever really goes out of its way to conflict with religion, but when you're making claims about things, you know, I talked about non-overlapping magisteria: the problem with that is that [its] almost never science [going] outside of its box, right? But the religious people, they keep going outside of that box and they'll make a testable claim either about the real world, in which case you can say, 'Well, no, that's just not the case, we have these reasons to conclude this' or they make odd claims about the supernatural, which you can say is outside of the realm of science, sure, but they'll make testable claims about the supernatural and so then any, kind of, shift in where the boundaries are themselves, then puts something into the realm of science."

Douglas saw two problems with the label of NOMA. He noted that religion frequently violates the principle of non-interference by making claims about the physical world, thus infringing upon the realm of science. However, he also saw a deeper issue in that the border that delineates science and religion is itself hard to define. He observed that some supernatural claims by religions can be tested empirically, and he does not want this to be seen as science infringing upon religion, but rather as an argument that the label of NOMA cannot be applied at all.

Of course, it is to be expected that someone who identifies with a 'hard' secular humanism will see intellectual problems with the 'softer' NOMA position. What was striking in our interviews, however, was how often the rejection of labels had little or nothing to do with intellectual questions, and instead related to the way in which such labels implied agreement with certain social movements or political arguments. Indeed, a common theme in our interviews was for participants to reject a label despite being partly, or entirely, aligned with it intellectually. Ronald, a Christian astronomer in Canada, accepted evolution and professed to hold a view that might be described as Theistic evolution, but he still eschewed calling himself an 'evolutionist' because that term is, he noted, fraught:

> "I don't consider myself an evolutionist because I never liked that word. It's always ... because I heard it used in such a pejorative way all the time. I don't do biology, so I don't know, but do I believe that it's a slam-dunk, that we ... came from less complex organisms...? Yes, of course."

In the case of 'New Atheism', too, we found similar wariness. Arduino is a professor of marine ecology who was born in Italy but now works in the UK. He is an atheist who sees religion as inimical to critical thinking, and who agrees with many claims of public figures characterized as members of the New Atheist movement. However, he distanced himself in interview from the label New Atheism because, he said, it implies a coherent group that is more dogmatic than atheist groups actually are:

> "I think it's sad in a way that we're talking about New Atheists because, first of all, the label 'New Atheist' has been given to them by someone else, so they don't proclaim themselves as New Atheists. Some of them are better at doing certain things than others, like, for example, I find

> Richard Dawkins a fantastic writer, a fantastic writer, I had read quite a few of his books even before he became a staunch anti-religious campaigner.... I think what is sad is that they have been building up New Atheists as if they were a new religion and often the New Atheist expression is being used as a label, as if it was another religion, but it works completely different. It is not a religion because it works using logic and rational thinking and religion is the antithesis of logic and rational thinking."

Moreover other atheist participants who strongly affirm a science–religion conflict rejected the label New Atheism because they, like the case of Dawkins mentioned earlier, did not see it as any different from earlier forms of atheism:

> "Like, New Atheism is a silly term because atheism isn't really a term, like, it's not a container into which you can pour meaning, really, it's just atheism. It's like new 'being dead' is the same as old 'being dead', right? So, in some ways, the distinctions of New Atheism are largely things that the media has given it." (Syd, musician, Canada)

Perhaps most striking, our UK sample incorporated a small number of individuals who could be described as creationists (in the sense that they rejected human and animal evolution entirely) but who personally disaffiliated from the label. Among the various shades of evolution rejection we encountered in the UK (but not in Canada, either due to national or sampling differences mentioned earlier), we only encountered one participant, a Seventh Day Adventist preacher, who identified, without prompting, as a creationist. All others had not heard of, were not interested in or actively disliked the concept of organized 'creationism'. This was the case for Lyle, for example, a radical Nonconformist Christian who rejected evolution but whose antipathy towards churches and other hierarchical religious bodies caused him to distrust institutionalized creationism as well.

### *Criticism and redefinition*

While in the quotes presented earlier, and in our sample overall, the rejection of labels dominated, in many cases, this was qualified by the individual interviewee. Some respondents observed that they would be prepared to accept labels if they were not connected in public discourse with normative stances they reject. Others felt constrained by what they perceived to be the narrow definitions of labels. In all these cases, how individuals constructed their own position was at odds with the pre-existing set of labels available in public discourse. Some of our respondents did not limit themselves to criticizing and rejecting labels, but went further in that they redefined them for their own purposes or selected some aspects that they deemed appropriate while leaving others. This was particularly common among the scientist interviewees, but was found across our sample. In this section, we provide and discuss examples of these forms of creative appropriation.

Some respondents arrived at their own label by rejecting other views. Oliver, a biologist at an elite university based in the UK, self-identifies as an evangelical, and although he broadly accepts evolution, confessed to having struggled with the subject due to his religious milieu. When asked how he would characterize his views on evolution and religion, he coined his own term based on a rejection of the other available labels: "[This is] a typically evangelical view, say what you're not.... Not a creationist, not Neo-Darwinist, sort of, blind evolution. I guess maybe Biblical evolution?".

Others did not create their own labels, but redefined available labels to fit their views, in many cases, by offering a literal reading of the label dissociated from its social connotations. For example, Walker, a Christian Canadian medical researcher, explained how he sees himself as a proponent of 'intelligent design', even though he rejects the ideas put forward by the intelligent design movement:

> "I'm very wary of [identifying with intelligent design]. So, on one level, I'll say, 'Yes, I believe in intelligent design', on one level. And by that, I mean if I believe that God created everything and I believe God is intelligent, then whatever he did was intelligently designed, right?"

Others performed a similar kind of redefinition with regard to creationism:

> "Yeah, so I'm a creationist, but as in: 'I believe God created the world but ... using the process of evolution to do that'. Yeah I guess that's ... as I've gotten heavily involved in evolution and stuff, yeah and my belief in Jesus hasn't really changed through that. So, I don't think that there has to be an issue there. So, for me, yeah, I hold firm to

> my faith and my study in biology. I mean many people ask me, like how or why, but, yeah, my interpretation I guess specifically of Genesis is, I don't have an issue with that.... I mean all Christians are creationists in the sense that they believe God created the world." (Greg, biology student, Christian, UK)

Walker and Greg used labels in a more general way than proponents of intelligent design and creationism frame them. Still others took this generalization even further and collapsed labels generally seen as distinct or even opposites. For example, Tina is a Canadian office worker. She teaches Yoga, and although she does not identify as Hindu, she is strongly committed to spiritual beliefs from that tradition, subscribing to the idea of karma, which she regards as validated by modern science. Her holistic spiritual approach means that she sees creation and evolution as two sides of the same coin:

> "Well, creationism and evolution have never really been in conflict with me.... In some levels, I think we're constantly, even if you believe in creationism, the Lord sat down for a week, had some time on his hands, made up a world for us to experience. Sure. Why not? It's no different, really, than a big bang. We would say the sound is the creation of the universe and the sound is Om. What do we say in the Bible? God spoke and created the world. And I've heard that they've actually sent deep space recordings out and that the sound that they record is the same frequency as the sound 'Om'."

Tina did not refer in her account of creationism to the specifics of Biblical literalism that are generally associated with that label. She generalized creationism to mean something akin to the universe being meaningfully structured, and she did not specify whether she believes this was done by God or another spiritual force. This level of generalization allowed her to see creation from nothing, as well as cosmic and biological evolution, as merely two possible ways of realizing the same spiritual insight.

Alongside these cases, other respondents handled labels creatively by combining different forms of generalization. Ronald, whose rejection of the label 'evolutionist' we already discussed, also rejected the label intelligent design, but he qualified this rejection using two forms of generalization:

> "Well, I'm certainly not a creationist; I have not been for a while, if I ever was. Intelligent design, let's face it, is just creationism 2.0. At least as [Michael] Behe and others marketed that idea … you were just a complexity, that kind of thing. I do believe in intelligent design in the sense that the universe, the author of the universe, the creator of the universe, the composer of the universe, the author of the universe, that there is intelligence in God's work. And so that's intelligent design. But I know what you mean by … at least I think what you mean by this term is the modern incarnation of creationism. At least a more sophisticated incarnation … but it is still creation[ism] nonetheless."

Ronald saw intelligent design as merely another, though veiled, form of creationism. In this, he mirrors arguments encountered earlier, though it is unclear if this influence is direct. Ronald's way of collapsing together labels that are kept apart by those who advocate and argue for them in public life is similar to Tina collapsing creationism and evolution. At the same time, Ronald generalizes what might be meant by intelligent design in a way similar to what Walker says about it. He believes in 'intelligent design' to the extent that he believes God is intelligent and created the universe.

## Conclusion

What is clear from our research is that the application of evolution and belief labels is fraught with difficulty, and for social scientists, carries a variety of risks. How, then, should sociological research into science and belief approach these deeply contested terms? Based on what we have seen here, we propose making a clearer distinction between two forms of analysis: first, interrogation of the language games found in public discourse about science and belief and people's everyday engagement with these games (Kaden et al, 2017); and, second, attempts to identify observable communities of belief and their ideological commitments. In both of these forms of analysis, labels can be validly used. In the first case, a central part of any analysis of public debates about science and belief will be how, and why, labels are used. In the second, any attempt to develop a typology of beliefs about evolution (or a related subject) found in a society will involve categorization and labelling. Problems emerge, however, when these two types of inquiry blur. There will, of course, be manifold interactions between systems of ideas that are publicly propagated and categories of popular belief (Kaden et al, 2017). If, however, technical labels developed in the context of public debates about science and belief are uncritically used to classify populations, social scientists risk not just creating misleading pictures (Elsdon-Baker, 2015), but unintentionally legitimating groups whose central aim is winning public recognition for a position on science and belief and any associated political objectives.

From what we have described earlier, some of the dangers of this should be clear. Not only did we find that people generally rejected the labels used in public discourse – with only a very small minority wholeheartedly embracing a label 'as is' – but they also frequently did this even in cases where their professed beliefs seemed to match up with a specific label. In many cases, this was because their alignment with the epistemic claims of certain creationist or anti-creationist organizations did not translate into alignment with their social and political goals. Both in the countries studied here and in the US (Evans, 2016), labels are associated with styles of public representation as well as with sets of ideas, and people can reject labels based on their perception of how those who are associated with them act in public and advance their case. Additionally, one question that social scientists need to ask is whether they should use categories that may contribute towards stereotyping. We described earlier several examples of participants who accepted evolution and who considered the universe to be created by an intelligent force. These usually felt constrained by the way in which the words 'intelligence' and 'creation' have, due to the activities of specific interest groups, acquired strong anti-evolution connotations.

Two proposals follow from this. The first is for the development of a different spectrum of labels that is less reliant on the terms of public debate. This could involve the creation of measures to assess the public identification with views about evolution (Hill, 2014) or the formulation of questions that bypass categorical definition in favour of questions that probe people's professed beliefs in a subtler way (Elsdon-Baker et al, 2017). This will mean, to reiterate a point that Jonathan Hill makes in this volume, placing more emphasis on meaningful categorization and less on replicating survey items. The second is for greater attentiveness to how publics *do* use labels, and, in particular, the creative ways in which they may be engaged or disengaged with. Again, there has been some encouraging recent research along these lines (Evans, 2016; Johnson et al, 2016). These changes could potentially move sociological research into science and belief into a different position: one of questioning, rather than reinforcing, the terms of debate.

## Notes

1 By 'public discourse', we mean popular media coverage of issues relating to science and belief and the publications of individuals that we have referred to elsewhere (Kaden et al, 2017) as 'professionals' in the field of science and religion whose vocation it is to develop, distribute, defend and critique systems of explanation that relate science to different forms of ultimate belief.

2 For further elaboration upon our use of 'science and belief', see our Introduction to this volume.

3 Prominent spokespersons for these labels, such as Stephen Jay Gould and Richard Dawkins, did disagree vigorously on aspects of evolution science, which is, of course, highly diverse itself, but these differences were largely restricted to specialist publications, in contrast to their arguments about religion.

4 See: www.reasons.org/

5 Intelligent design was subsequently ruled a form of creationism in Kitzmiller v Dover Area [Pennsylvania] School District, a federal district court case in 2005.

6 UK interviews and focus groups were conducted predominantly in the North-west and West Midlands of England. The majority of the Canadian interviews were conducted in Ontario.

7 Needless to say, this division is crude and masks the substantial differences in our sample in terms of levels of practice, strength of belief and manner of identification.

8 All participants have been assigned pseudonyms, and we are very grateful to them for volunteering to participate in the research.

9 Julie was the only participant to do this.

## Acknowledgements

The authors would like to extend their thanks to Tom Aechtner and Chris Toumey for their comments on an earlier draft of this chapter.

## References

Aechtner, T. (2016) 'Challenging the Darwin skeptics: examining proevolutionist media persuasion', *Journal of Media and Religion*, 15(2): 78–99.

Answers in Genesis (no date[a]) 'Young-earth creationism'. Available at: https://answersingenesis.org/creationism/young-earth/ (accessed 22 November 2017).

Answers in Genesis (no date[b]) 'Intelligent design'. Available at: https://answersingenesis.org/intelligent-design/ (accessed 22 November 2017).

Behe, M.J. (2006) *Darwin's black box: The biochemical challenge to evolution* (2nd edn), New York, NY: Free Press.

Bishop, G. (2003) '"Intelligent design": illusions of an informed public', *Public Perspective*, 14(3): 40–2.

Campbell, J.L., Quincy, C., Osserman, J. and Pedersen, O.K. (2013) 'Coding in-depth semistructured interviews: problems of unitization and intercoder reliability and agreement', *Sociological Methods and Research*, 42(3): 294–320.

Dawkins, R. (1986) *The blind watchmaker*, New York, NY: Norton.

Dawkins, R. (2016) *The god delusion; 10th anniversary edition*, London: Black Swan.

Elsdon-Baker, F. (2015) 'Creating creationists: the influence of "issues framing" on our understanding of public perceptions of clash narratives between evolutionary science and belief', *Public Understanding of Science*, 24(4): 422–39.

Elsdon-Baker, F., Leicht, C., Mason-Wilkes, W., Preece, E. and Piggot, L. (2017) *Science and religion: Exploring the spectrum: Summary report of preliminary findings for a survey of public perspectives on evolution and the relationship between evolutionary science and religion*, Birmingham: Newman University and YouGov.

Emmel, N. (2013) *Sampling and choosing cases in qualitative research* (1st edn), London: Sage Publications Ltd.

Evans, J.H. (2011) 'Epistemological and moral conflict between religion and science', *Journal for the Scientific Study of Religion*, 50(4): 707–27.

Evans, J.H. (2018) *Morals not knowledge: Recasting the contemporary US conflict between religion and science*, Oakland, CA: University of California Press.

Evans, M.S. (2016) *Seeking good debate: Religion, science, and conflict in American public life*, Oakland, CA: University of California Press.

Gardham, D. (2008) 'Richard Dawkins: Muslim parents "import creationism" into schools', *The Telegraph*, 3 August. Available at: www.telegraph.co.uk/news/uknews/2494397/Richard-Dawkins-Muslim-parents-import-creationism-into-schools.html (accessed 10 April 2018).

Gieryn, T.F. (1999) *Cultural boundaries of science: Credibility on the line* (2nd edn), Chicago: University of Chicago Press.

Gould, S.J. (2002) *Rocks of ages: Science and religion in the fullness of life*, Cambridge: Ballantine Books.

Hill, J.P. (2014) *National study of religion and human origins*, Grand Rapids, MI: BioLogos Foundation.

Johnson, D.R., Ecklund, E.H. and Matthews, K.R.W. (2016) 'Responding to Richard: celebrity and (mis)representation of science', *Public Understanding of Science*, 27(5): 535–49.

Kaden, T., Jones, S., Catto, R. and Elsdon-Baker, F. (2017) 'Knowledge as explanandum: disentangling lay and professional perspectives on science and religion', *Studies in Religion*, first published online, 22 December. Available at: https://doi.org/10.1177/0008429817741448

Kettell, S. (2013) 'Faithless: the politics of new atheism', *Secularism and Nonreligion*, 2: 61–72. Available at: www.secularismandnonreligion.org/articles/17/

Kuebler, D. (2016) 'What's in a name? Does Darwin hinder the acceptance of evolution?', Science and Religion: Exploring the Spectrum Blog. Available at: https://sciencereligionspectrum.org/blog-posts/whats-in-a-name-does-darwin-hinder-the-acceptance-of-evolution/ (accessed 10 April 2018).

LeDrew, S. (2016) *The evolution of atheism: The politics of a modern movement*, New York, NY: Oxford University Press.

Long, D.E. (2011) *Evolution and religion in American education: An ethnography*, Dordrecht: Springer.

Melott, A.L. (2002) 'Intelligent design is creationism in a cheap tuxedo', *Physics Today*, 55(6): 48–50.

Meyer, S.C. (2010) *Signature in the cell: DNA and the evidence for intelligent design*, New York, NY: HarperOne.

Numbers, R.L. (2006) *The creationists: From scientific creationism to intelligent design* (exp edn), Cambridge, MA: Harvard University Press.

Pennock, R.T. (eds) (2001) *Intelligent design creationism and its critics: Philosophical, theological, and scientific perspectives*, Cambridge, MA: MIT Press.

Ramm, B. (1981 [1954]) *The Christian view of science and scripture*, Grand Rapids, MI: William B. Eerdmans.

Scott, E.C. (2009) *Evolution vs. creationism: An introduction* (2nd edn), Berkeley, CA: University of California Press.

Scott, E. and Branch, G. (2006) *Not in our classrooms: Why intelligent design is wrong for our schools* (1st edn), Boston, MA: Beacon Press.

Toumey, C.P. (1994) *God's own scientists: Creationists in a secular world*, New Brunswick, NJ: Rutgers University Press.

Wengraf, T. (2009) *Qualitative research interviewing: Biographic narrative and semi-structured methods* (1st edn), London: Sage Publications Ltd.

4

# Researching Clergy Attitudes towards Science: A Reflective Account of Key Methodological Challenges

*Lydia Reid*

## Introduction

As a sociologist carrying out research on clergy attitudes towards science, one of my favourite tales concerns a senior church leader who volunteered to be interviewed. When we met up, I thanked him for agreeing to speak to me about Christianity and science. At this point, the colour began to drain from his face: 'Oh dear!', he said, 'I thought you were here to talk to me about Christianity and *silence*!'. While this is a light-hearted story, the statement that followed gave me a real insight into one of the key challenges of doing research in this area. The interviewee went on to say: 'If I had realized you were interviewing me about science, I probably wouldn't have agreed to do it'. When I probed him further, he replied that not being from a science background would have put him off coming forward to be interviewed. I had suspected, anecdotally at least, that this might be the case since science is often perceived to be a specialized subject that only an 'expert' can comment on. As Katherine Mathieson (2017), chief executive of the British Science Association, observes: '[S]cience needs to be taken out of its cultural ghetto.... [I]t's seen as the realm of professionals and experts'. Mathieson goes on to say that science is guiltier of this than other academic/societal spheres such as business, arts, politics or sports. In the case of the church leaders in my research, this apprehension around science also fed into a general mistrust of the media, with many of my interviewees citing their own experiences of having taken part in intentionally divisive radio or TV panels.

The aim of this chapter is to outline some methodological reflections derived from my own experiences of carrying out research on science and Christianity. I want to argue that being more reflexive about one's own methodological challenges can lead to the refinement of research methods and enable greater understanding of research subjects and the social world. This is particularly needed in the case of science and religion research, which is relatively undeveloped. Indeed, social-scientific research on this topic has only recently begun to flourish in the UK and Europe; this collection thus contributes to an under-researched field. In the US, research into science and religion has been more prevalent, and the works of Hill (2014, 2015), Evans (2011; see also Evans and Evans, 2008), Ecklund (2010; see also Ecklund et al, 2011), Scheitle (2011; see also Scheitle and Cornell, 2015; Scheitle and Ecklund, 2017), Baker (2012) and others continue to be influential.

Given that the establishing of a social-scientific approach to studying science and religion is only in its early stages (beyond the US at least), there will inevitably be some unique challenges along the way. Not least among these is the fact that – as Jones et al observed in the Introduction to this volume, in reference to Evans and Evans's (2008) work – the discipline to which I belong, sociology, is part of science and religion's complex history. Indeed, sociology's own inception (through the writings of Auguste Comte) reinforced the conflict narrative (an area I will expand on later in the chapter) (Evans, 2011: 708). One useful way of visualizing this is to think about parallel mirrors continuously reflecting on each other: one mirror as the tool of study, sociology; the other as the research area of science and religion. Two key figures who have inspired my own methodological thinking in this area are Elsdon-Baker (2015; see also Elsdon-Baker and Mason-Wilkes, Chapter 1, this volume) and her research on 'creating creationists' in survey frameworks, and Hill's (2015; see also Hill, Chapter 2, this volume) analysis of whether Americans believe that science and religion are in conflict. I will reflect upon both in this chapter. The key themes that I will explore are: the self-selecting nature of research respondents; the problem of using pre-existing typologies (open versus closed questions); and issues in identifying and understanding creationist positions.

## Project and methods

The research presented in this chapter forms part of the 'Equipping Christian Leadership in an Age of Science' (ECLAS) project funded by the Templeton World Charity Foundation. The project is headed by Canon Professor David Wilkinson (cosmologist and theologian), Professor Tom McLeish (physicist and Anglican lay reader), Revd Dr Kathryn Pritchard (research fellow/project manager who works with the national Mission and Public Affairs Division for the Church of England [C of E]) and the Rt Revd Dr Richard Cheetham (Bishop of Kingston and former science teacher). The project is multi-stranded and aims to do the following:

- provide theological training resources for ministers (common awards);
- fund 'Scientists in Congregations' schemes (bringing together clergy and scientists from within the congregation to formulate events/research in order to contribute to the science and religion dialogue);
- host conferences on key scientific areas in conjunction with the academic scientists based at Durham University (covering topics such as cosmology, the environment, evolution, neuroscience, end of life and artificial intelligence);
- contribute an original piece of research on clergy attitudes towards science; and
- provide expert advice/support on complex issues (achieved through having a presence in the Archbishops' Council, Church House).

The project is theologically inclined and, as such, has a commitment to supporting and facilitating the dialogue between science and Christianity. It is an unusual project in bringing together academic scientists (from physics and cosmology), theologians and myself, a social scientist.

The research methodology consisted of one-to-one qualitative interviews (chiefly with senior church leaders) and a quantitative survey of over 1,000 clergy. The project is ecumenical and includes responses from Anglican, Methodist, Baptist, United Reformed Church (URC), Pentecostal and Catholic clergy. However, due to the project's partnership with the C of E, and our reliance on its gatekeepers (an area that I will expand on later in the chapter), a larger proportion of our sample has turned out to be Anglican.

## 'Frustrated access' and the self-selecting nature of research participants

Research into clergy attitudes towards science remains a neglected area, and where such research has been carried out, it has generally included very small sample numbers, often recruited through convenience methods. Four examples of studies focusing exclusively on clergy to which I will refer throughout this chapter have been provided by Dickerson et al (2008), Colburn and Henriques (2006), Bouveng and Wilkinson (2016) and Gregory (2017). As Village and Baker (2013: 558) point out: '[C]ommitted churchgoers are relatively scarce in nationally representative UK samples because only around half the population now affiliates to religion'. The number of *clergy* in the UK is, of course, significantly less. There are, then, inherent difficulties in carrying out research with this very specific (and small) subset of the population, and these are compounded when part of this subset is, as I found, reluctant to participate. Rather than ignoring such difficulties, however, sociologists such as Gilliat-Ray (2005) encourage researchers to see access problems as being part of the data themselves. Concerning her own research on Islamic colleges, she writes:

> [I hope] my writing about (non) access will encourage scholars to similarly share their 'stories' of denied or frustrated access. By not hearing accounts of failed research (which, of course, are not usually reported!) we are denied a sense of the field as a whole, and the way in which it might be 'skewed' by reliance upon studies which have been undertaken and which have not presented insurmountable difficulties. (Gilliat-Ray, 2005: 29–30)

Gilliat-Ray's use of the phrase 'frustrated access' sums up the challenges that I and my predecessor, Dr Rebecca Bouveng (who worked on the preliminary stage of the research for ECLAS), faced when carrying out research with clergy. Bouveng contacted over 30 senior church leaders from a variety of Christian backgrounds, securing 14 interviews in total, with two cancelling last minute. In her article written with Wilkinson she notes:

> It is significant that only sixteen [senior church leaders] accepted the invitation [to be interviewed on science]. Some declined for purely practical reasons, but it is clear from some replies that some did not feel confident enough

> to participate in this anonymous academic research project. Two bishops were explicit in declining on the grounds that they did not know enough, while another senior leader asked to have two of his advisors at the interview. (Bouveng and Wilkinson, 2016: 101–2)

Here, Bouveng and Wilkinson identify a variety of potential reasons why church leaders may decline an interview on science. My own experience supports the idea that church leaders often describe themselves as being 'too busy' to take part or simply do not feel that they are qualified to talk on the topic. One bishop declined on the grounds that he was '[E]ducated in the age of the two cultures', a reference to C.P. Snow's 1959 lecture, where he draws a distinction between 'literary intellectuals' and 'scientists' (Snow, 2012: 4). Snow was making the point that neither group should remain separate, but it has since become shorthand for an allegedly growing divide between the humanities and natural sciences. Presumably, the bishop saw himself as belonging to the humanities and, moreover, that this was a suitable justification for his non-participation in the interview. There were two further instances where bishops made this distinction (one during an interview and the other at a conference) and one might speculate whether this became an accepted part of the cultural milieu for students/ministers training in the 1950s and 1960s. These issues were also reflected at a session at the College of Bishops, where the ECLAS project analysed the responses of bishops when asked about their views on talking publicly about science. Three strategies emerged: 'silence' (knowing when it is more appropriate to stay silent on a science issue); 'delegation' (handing over a science topic to another senior church leader who is perceived to know more in the area); and 'engagement' (speaking out, which can be positive or negative depending on the experience). I wondered to what extent this 'silence' strategy may be playing out in the low numbers of bishops agreeing to be interviewed on science. Moreover, I encountered 'delegation' in a number of emails, where bishops said that they were not qualified to discuss science, but could point me in the direction of someone that could.

In recruiting interviewees, I emailed over 40 senior church leaders, gaining 20 interviews in total. For non-Anglican Church leaders, relevant gatekeepers were asked to circulate an invitation on my behalf. The breakdown of participants is outlined in Table 4.1.

The most difficult group to secure were the C of E bishops.[1] While a proportion of these responded by saying that they were 'too busy'

**Table 4.1: Christian affiliation of church leaders interviewed**

| C of E | Catholic | Baptist | URC | Methodist | Pentecostal |
|---|---|---|---|---|---|
| 9 | 2 | 3 | 3 | 2 | 1 |

to take part, I also suspected that my being outside of C of E circles impacted on bishops' readiness to decline the interview. It is interesting that when I worked with contacts in the Mission and Public Affairs Division of the Archbishops' Council, I was able to secure interviews with a further four Anglican bishops. Both with the interviews and the survey (which I will refer to shortly), there was a sense that the project's importance needed to be 'vouched for' by trusted figures in the C of E. This is not unusual, and Crano and Brewer (2002: 237) rightly point out that the gatekeeper 'can affect the likelihood of a respondent's cooperating with the interviewer'.

Another variable to consider is the number of interviewees who came from a 'science background' (by which I mean an individual who has studied a science, technology, engineering and/or maths (STEM) subject at degree level or higher). Of particular concern here – and the flip side of 'frustrated access' – is the possibility that those who agree to take part in the interviews are self-selecting and, as Bouveng and Wilkinson (2016: 101–2) rightly point out, skewed towards 'a sample of leaders who themselves do not fear engagement with science'. This was found in a study by Dickerson et al (2008), who note that in their sample of 63 United Methodist Ministers, just under half had a degree or higher in science and 94% rated their science knowledge as 'good' or 'very good'. Dickerson et al (2008: 368) refer to this latter percentage as 'troubling' since it may indicate 'overconfidence in their understandings'. Self-selection can therefore be influenced by science background and perceived knowledge of science. In my own research, of the 20 participants I interviewed, only five came from what I might consider a 'science background'. From a methodological point of view, this is somewhat encouraging as it challenges the idea that *only* those from a science background would be willing to be interviewed on science. This finding also indicates the variation in how church leaders view science, from the interviewee mentioned at the beginning of this chapter (who would not have agreed to be interviewed had he known the topic was on science), to the remaining 13 participants who were from a non-science background but were still willing to talk about the subject.

One important discovery that I made during the interviews, however, was the influence of close friends or family members who were scientists. Those interviewees who were not scientists themselves often spoke enthusiastically about friends or family who worked or studied in science. Out of the 20 senior church leaders included in the sample (minus the five that were from science backgrounds themselves) a further five reported having someone in their immediate family who was a scientist. With this in mind, half my interview sample had some kind of personal or social connection to science. Such a finding raises the possibility that participants' eagerness to take part in the interviews might also depend on the extent to which science is 'normalized' in their respective social and/or familial networks. The role of social and family networks is well established in research on science engagement, as well as on science and religion. Research carried out by Archer et al (2012: 884) on the aspirations of children towards science points towards the influence of 'family habitus' on making science aspirations for teenagers more 'thinkable'. Similarly, in his research on the persistence of anti-evolution stances over time among young people, Hill (2014: 575) concludes that 'social networks play an important moderating role' and are more important than educational attainment. Scheitle and Ecklund's (2017) exploration of religious parents' likelihood of recommending a STEM career also found that a low level of science interest was the main indicator in parents not recommending STEM careers. Given this, it is possible that social networks *and* personal interest influence how science is perceived by clergy. Indeed, the remaining 10 of my interviewees (who were not scientists and did not report a scientist in their social networks) did talk about science enthusiastically and with a sense of personal interest. This latter point was also found to be an important factor in research conducted by Scheitle and Cornell (2015), who found that personal interest in a topic meant that congregation members were more likely to remember it being preached about by their church leaders.

There were also indications of the role of family networks in the quantitative element of the survey (which I will refer to later in the chapter). For the survey, I was unable to draw upon pre-existing data sets and decided not to administer the survey through a research company (due to the sample being too specific). Some responses from Baptist and Pentecostal groups were elicited through email invitations, with these responses likely being subject to similar self-selection. Most responses, however, were collected using paper questionnaires at diocesan clergy days (usually during the coffee or lunch break). On some occasions, I was given permission to complete the survey as part of the schedule of the day. In one notable example, 250 clergy filled in the questionnaire at one event. While there was still a small element of self-selection in this process (in that the survey was not mandatory), most respondents felt obliged to fill in the questionnaire since it had been vouched for by the organizers. The target for responses had been set at 1,000 and, as with the interviews, a point began to be reached (around 400 responses) where securing permission for clergy days became more difficult. To counter this, the project team began offering prospective respondents a chance to win a raffle prize of £2,000 for a church of their choice. The advantage of having the monetary incentive was that I was more likely to include clergy who might be indifferent or unenthusiastic about science. Interestingly, this resulted not only in more clergy being willing to fill in the questionnaire, but also in a greater variety in the quantity and quality of what was written. One of the data-entry clerks commented on the disparity between those clergy who wrote several paragraphs on open-ended questions and those who wrote one-word answers. While researchers typically consider people's answer to a survey question as the most important measure of data, the *way* in which they answer (eg in quantity) also provides us with cues for how respondents *feel* about filling in a questionnaire on a particular topic, as I will explore later.

Working in conjunction with the consultant statistician to the project, Dr Tim Drye, I devised a questionnaire that explored the extent to which clergy are, or are not, engaged with science-related issues. The questions explored science issues in relation to personal conversations, independent research and public/media figures. The aim of this three-tiered approach was to explore the nature of scientific engagement in a more dynamic way. The questions were primarily closed in nature; however, staying faithful to the qualitative side of the project, and with the aim of introducing more 'creativity' in survey questions (see Hill, Chapter 2: this volume: 45–6), I did incorporate some open-ended questions that allowed respondents to reflect on their engagement with science and what they had learned. The presence of a scientist in families or friendship groups was so pertinent in the interviews that it went on to inform an additional survey question on that subject (see Table 4.2). Interestingly, this question revealed a statistically significant[2] difference between those who reported themselves as having a close friend or family member who was a scientist and their likelihood of 'disagreeing' or 'strongly disagreeing' with the statement that 'The money spent on sending a spacecraft to Comet 67P should have been spent on providing clean water for people throughout the world'. The 'strongly agree' and 'agree' columns for the Comet 67P statement are predominantly made up of those who do *not* have members of family or close friends who

**Table 4.2: Cross-tabulation on family/friend from science background and money spent on Comet 67P**

| | | The money spent on sending a spacecraft to Comet 67P should have been spent on providing clean water for people throughout the world | | | | | |
|---|---|---|---|---|---|---|---|
| | | Strongly agree (%) | Agree (%) | Neither agree nor disagree (%) | Disagree (%) | Strongly disagree (%) | Total (%) |
| Do you have any family members or close friends who are from a science background? | Yes | 7.1 | 24.6 | 35.8 | 28.9 | 3.5 | 100.0 |
| | No | 11.1 | 27.9 | 33.9 | 24.7 | 2.4 | 100.0 |
| Total | | 8.5 | 25.8 | 35.1 | 27.4 | 3.2 | 100.0 |

are scientists, while those at the other end who 'strongly disagree' and 'disagree' are more prevalent among those who do have scientists in their family/social networks. It is important to note, however, that there is still a sizeable chunk in the middle expressing hesitation about the comet statement regardless of whether or not they have family members/close friends who are from a science background.

## Pre-existing typologies (open versus closed questions)

At the beginning of the chapter, I referenced the work of Hill (2015; see also Hill, Chapter 2, this volume) and his research on whether or not Americans believe that science and religion are in conflict. Hill's work reminds social scientists to think critically in their design and implementation of survey questions. Commenting on a question about conflict in the Pew Research Center's survey (which gives two options: that science and religion are 'often in conflict' or 'mostly compatible'), he notes:

> Variations on this question produce quite different results. If, instead of two possible positions, the respondent is asked to pick from *three* possible positions – that science and religion 'generally agree', 'generally conflict' or are 'not related to each other in any meaningful way' – then only between one quarter and one third pick the conflict response. (Hill, Chapter 2, this volume: 27)

This flags up one of the major problems of doing survey research on science and Christianity, namely, that the researcher is unable to fully capture the nuance and context that can otherwise be found in qualitative interviews. Given that my sample was clergy, I took an active interest in theologians' conceptions of science and religion, thereby breaking with the norm that, in Evans and Evans's (2008: 90) words, theological literature remains 'outside of the interests of most sociologists'. There are a number of well-known theologians who promote a broadly dialogic or integrative approach to science and Christianity, for example, McGrath (2016), Polkinghorne (1996) and Barbour (2000). According to Cantor and Kenny (2001: 765), Barbour is 'probably the most widely cited author in the area of science and religion', with his books offering a 'sophisticated overview of the field'. For this reason, I selected Barbour's typology on science and religion to use in the survey. His typology includes four ways in which science and religion can relate to each other – 'conflict', 'independence', 'dialogue' and 'integration' (Barbour, 2000: 2–3) – and using his framework in research allows us to see if/how these categories 'map' onto people's actual beliefs (see also Kaden et al, Chapter 3, this volume).

'Conflict', as one might infer, presents science and religion as two antagonistic spheres, sometimes known as the 'warfare' model. Barbour uses the example of 'Biblical literalists' as a group of believers who would see evolutionary science as being in conflict with their own Christian beliefs. He also refers to 'atheistic scientists', who claim that evolutionary science cannot be accepted by theists (Barbour, 2000: 2). This, of course, is a crude generalization but it presents science and religion as two spheres that are interactive, competitive and both trying to assert truth claims in the public square.

'Independence', like 'conflict', presents science and religion as separate entities; however, they differ in their terms of engagement. Conflict requires interaction whereas independence does not. In effect, there is a 'compartmentalization' of science and religion that avoids conflict but so, too, any constructive interaction (Barbour, 2000: 2). Independence can represent two slightly different positions on science and religion. The first is that both spheres are asking different questions about the world; Bouveng and Wilkinson (2016), for example, found in the accounts of their interviewees that the how–why distinction became an attractive way for church leaders to conceptualize science and Christianity. This meant that science was responsible for the 'how' questions (eg how things work, the use of objective facts and empiricism) whereas Christianity was responsible for the 'why' questions (eg why we're here, moral values and the derivation of meaning) (Bouveng and Wilkinson, 2016: 106–7). Interestingly, the group of leaders I interviewed did not make reference to this distinction; instead, they tended to emphasize the commonalities between science and religion in terms of both domains seeking truth. However, it *was* a more popular description in the survey responses (an area that I will expand on shortly). Moreover, Bouveng directly asked the senior church leaders she interviewed whether they agreed with the how–why distinction. This shows that introducing concepts can 'force' answers that might otherwise not occur organically. The second position found under 'independence' states that science and religion can offer complementary perspectives on the world but are not mutually exclusive. In other words, science and religion can complement one another but they still remain in separate spheres.

'Dialogue' entails science and religion being in conversation with each other. There is interaction between the two spheres but, unlike 'conflict' and 'integration', there is a clear sense of belonging to separate domains. As Barbour (2000: 3) states: '[B]oth scientists and theologians are engaged as dialogue partners in critical reflection … while respecting the integrity of each other's fields'. It is this mention of 'respect' and 'integrity' that makes it different to 'conflict'. Barbour also points out that dialogue can form a comparison between the two fields, showing both similarities and differences. Alternatively, dialogue may happen when science raises 'limit questions' that it cannot answer, such as 'Why is the universe orderly and intelligible?' (Barbour, 2000: 3). A third form of dialogue centres around the communication of information, with concepts from science being used as 'analogies for talking about God's relation to the world' (Barbour, 2000: 3).

'Integration', one can see, is Barbour's own preferred position (along with dialogue), for, as he notes, this interaction is 'a more systematic and extensive kind of partnership between science and religion' (Barbour, 2000: 3). It is arguable that the ECLAS project on which I work is, in its overarching ethos, seeking to take a more integrated approach to science and religion. However, such integration cannot take place without dialogue.

Barbour's typology is useful in that it provides a more detailed framework than some of the other options presented in survey questions on science and religion. However, the types are still unclear at times. Take the 'independence' model, for example; another key advocate of this approach was Stephen Jay Gould (1997). Gould proposed that science and religion should be characterized as 'non-overlapping magisteria' (NOMA), a position he developed as a reaction, in part, to the growing creationist movement in America. In his original writing on NOMA, he states:

> These two magisteria [science and religion] do not overlap, nor do they encompass all inquiry. To cite the arch clichés, we get the age of rocks, and religion retains the rock of ages; we study how the heavens go, and they determine how to go to heaven. This resolution might remain all neat and clean if the nonoverlapping magisteria (NOMA) of science and religion were separated by an extensive no man's land. But, in fact, the two magisteria bump right up against each other, inter-digitating in wondrously complex ways along their joint border. (Gould, 1997: 16)

What becomes apparent from this quote is that while separation is the defining feature of science and religion, there does come a point at which 'the two magisteria bump right up against each other', interacting on their 'joint border' (Gould, 1997: 16). This latter point seems counter-intuitive because NOMA, at first glance, suggests *no interaction*. Gould's biographers have also noted that he 'advocated that the relationship between science and religion should be characterized by "intense dialogue"' (Kelley, 2009: 175). Despite the fact that both domains have their own tools for making sense of the world, with science dealing with empiricism and religion dealing with morality, the two are meant to interact and be in dialogue. The very fact that they are interacting means that they are participating in a shared reality where both science and religion seemingly have something to contribute. NOMA, while useful, could end up being used as shorthand for the total separation of science and religion, with no dialogue between the two. It is clear from both Gould and Barbour that the relationship between science and religion is complex, nuanced and multilayered, and that there are inconsistencies between and within typologies of the relationship. The use of typologies as research instruments is arguably flawed; however, they can highlight revealing gaps between philosophical discourse and everyday belief, as well as helping to improve or refine our techniques as researchers.

Despite the normative dimension of Barbour's typology, he is a widely recognized authority on science and religion, particularly in theological circles, as acknowledged by Colburn and Henriques (2006). With this in mind, his framework is likely to be known to the clergy I surveyed. The Barbour typology survey question that I used in the survey is shown in Figure 4.1.

**Figure 4.1: Barbour typology survey question**

Which of the following diagrams most closely reflects your own views on how science and Christianity *should* relate to one another? (Please circle a, b, c or d)

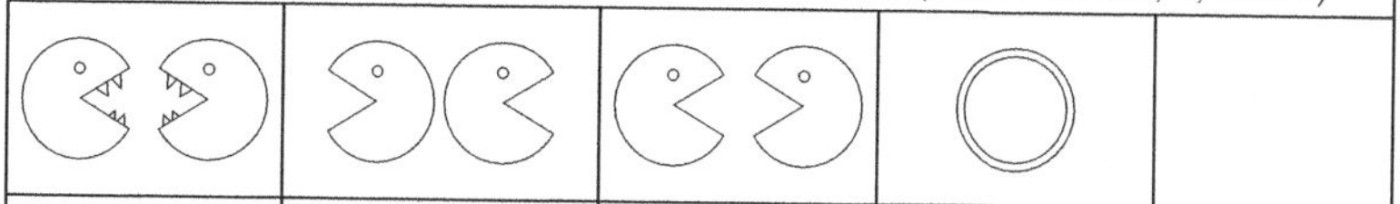

| a) Conflict | b) Independence | c) Dialogue | d) Integration | e) Other |
|---|---|---|---|---|

The word '*should*' rather than '*do*' in the survey question was purposely chosen: by using 'should', I am asking the respondent to indicate their own normative position on the relationship between science and Christianity. While this may result in an overstatement of non-conflict positions, it does tell us something pertinent about how clergy view the relationship between science and Christianity at an *aspirational* level, rather than in purely descriptive terms. Of course, this question provides insight into only one dimension of clergy attitudes towards science and religion (though I included an open-ended follow-up question asking the respondent to explain their rationale). It thus contrasts with surveys that have made a distinction between the public and personal levels in the conflict between science and religion. For example, the 2014 Pew Research Centre survey mentioned earlier in the chapter finds that when US adults are asked at a general level whether they think science and religion are 'often in conflict' or 'mostly compatible', 59% of respondents affirm the former and 38% the latter. When asked whether science 'sometimes conflicts' with *their own* religious beliefs, however, 30% say that it conflicts whereas 68% say that it does not (Funk and Alper, 2015; for other examples of this contrast, see Baker, 2012; Ecklund et al, 2011; Elsdon-Baker et al, 2017; see also the review by Hill, Chapter 2, this volume). This marked variation suggests that the phrasing of questions can be revealing – whether one is exploring public, personal or normative dimensions – with each touching on subtly different themes.

Of the clergy I surveyed, I found that 'integration' was the most common answer (51%), closely followed by 'dialogue' (45%), 'other' (4%) and 'independence' (0.4%). Perhaps unsurprisingly, 'conflict' was rarely chosen (0.2%), though one can reasonably presume that this would have been different had the survey asked clergy to comment on *public* perceptions of science and religion. From this question alone, one might be tempted to make an overly simplistic conclusion along the lines of: clergy in England overwhelmingly do not see science and Christianity as being in conflict. However, what the open-ended question (see Table 4.3) shows is that despite respondents ticking one answer, they then follow up their choice with an extended and multifaceted account of how science and religion interact. The extracts presented were chosen on the basis of showing this complexity across the different typologies. The word 'conflict' is used by two respondents who had ticked 'dialogue' and 'other', but softer synonyms, such as 'challenge', were more commonly used. The word 'challenge' appears in the accounts both of those who had ticked 'integration' and of those who had ticked 'dialogue'. David Gregory asked a similar typology question in his small-scale preliminary survey with Baptist ministers. Using Berger's (1970) model ('challenge to faith', 'irrelevance', 'separate', 'complementary' and 'enhancing'), he found that 'among ministers the vast majority see science and faith as either complementary or enhancing' and, similarly to my own research, that 'a quarter of respondents also stated it was a challenge to faith' (Gregory, 2017: 6). Bouveng and Wilkinson (2016: 103), in their research

**Table 4.3: Examples of responses to Barbour survey question**

| Response ticked | Open-ended question response: 'What are your reasons for your chosen answer?' |
|---|---|
| Integration | I see no conflict between faith and a scientific explanation of the universe, Science investigates how, Faith explores why. |
| Integration | Need to explore developments and challenges. |
| Integration | Science should not try to replace God. |
| Integration | I still disagree with natural science, it was faith that helped in the face of suffering not science, science provided info are both totally integrated [sic]. |
| Integration | Integration sounds like a counsel of perfection, and yet integration has to be the ultimate goal, in as much as it is the point towards which the universe is heading in God's purposes. Of course, dialogue and integration are not mutually exclusive, and as we move towards integration, dialogue will be essential. |
| Dialogue | I believe Christians should be unafraid to express what they believe and secure enough in it to have it challenged by science. In the end I believe they compliment one another [sic]. |
| Dialogue | They answer different questions, 'how' and 'why', but also has conflicts [sic]. |
| Dialogue | Integration may be the goal, but the starting points are different and dialogue is the path we need to travel. |
| Other | I think it has all aspects, conflict say on ethical grounds, independence as don't always sit together. |
| Other | Overlapping circles. I think some science is dishonouring to God, but most is awesome! |

with senior church leaders, also found that although most of their interviewees emphasized compatibility, mutuality or complementarity among science and Christianity, they did also reference 'tensions'. These tensions typically emerged in popular discourse, in the ethical, economic and political questions arising from science application, and in the implications of science for personal Christian faith.

Another significant point to emerge from this sample of responses is how the conflict thesis is still played out in the qualitative descriptions – even if the word is not used. Two of the 'integration' responses were followed by the statements 'Science should not try to replace God' and 'I still disagree with natural science, it was faith that helped in the face of suffering not science'. Under the 'other' category one respondent put: 'I think some science is dishonouring to God'. All three examples are highly critical of (some) science despite the first two respondents ticking 'integration' and the final one ticking the 'other' option and referring to 'overlapping circles'. What becomes apparent from the examples discussed earlier is that while 'conflict' is very rarely cited as being the *defining* feature of how science and Christianity should relate to one another, it still potentially features as part of the backdrop, even for those who have ticked 'integration'. This supports a comment made by Hill, when reflecting on why some people might see science and religion as being in conflict but not necessarily incompatible: he notes that, 'For many … the relationship between religion and science is probably not one of general epistemological conflict, but one of conflict over very particular knowledge claims' (Hill, Chapter 2, this volume: 30). Moreover, strongly divergent qualitative answers end up being part of the same Barbour typology. For example, the how–why distinction, which I referenced in relation to Gould's NOMA framework, also appears among respondents ticking 'dialogue' *and* 'integration'.

## Identifying and understanding creationist positions

Another area where confusions can arise is in identifying and understanding creationist positions and anti-evolution stances. In Elsdon-Baker's (2015: 424) research on 'creating creationists', she notes that the way in which researchers frame polls on religion and evolution might be creating a distorted or exaggerated picture of the conflict between the two. Elsdon-Baker reviewed the framing of four major surveys (one of which she developed herself) on public perceptions of whether or not there is a clash between evolutionary science and believing in a deity. The first notable point about the surveys included in Elsdon-Baker's research is that they were selected 'because they were in part, if not wholly, designed to be used within media campaigns' (Elsdon-Baker, 2015: 424). As I pointed out earlier in this chapter, the media often have their own agenda when reporting on issues like science and Christianity since their primary focus is on provocative headlines and attracting viewers/readers. One stark example of this can be found in Elsdon-Baker's discussion of a poll used in the BBC TV documentary 'Horizon: a war on science'. Elsdon-Baker found that the choice of answers to the evolution question was highly constraining (see Kaden et al, Chapter 3, this volume). In particular, the 'evolutionary theory' option explicitly made reference to God having no part in the evolutionary process, thereby excluding respondents who may take a broadly deistic or theistic evolutionary approach. The survey question itself sets up the clash between science and Christianity by equating evolutionary stances with an agnostic or atheistic position. However, as Elsdon-Baker notes, it is not just media polls that set up this distinction. She observes that there is 'a wider philosophical concern over markedly "atheistic" stances employed in issue framing in public opinion polls … and scholarly representation of a clash narrative between "religion" and "evolution"' (Elsdon-Baker, 2015: 423).

Informed by Elsdon-Baker's reservations about the framing of questions on evolution, I will now reflect on my own research with clergy into their views on this subject. In response to the statement '[B]iological evolution is the best model we have for understanding the natural world', 25% of the clergy I surveyed 'neither agreed nor disagreed', outnumbering those who 'strongly agreed' (18.4%), 'disagreed' (12%) and 'strongly disagreed' (4.8%). Only those who 'agreed' formed a larger group (39.7%). Interestingly, the clergy that ticked either 'agree', 'disagree' or 'neither agree nor disagree' had an almost equal number of people in each category who had previously selected 'integration' or 'dialogue' under the Barbour typology question. The only groups displaying a noticeable difference were those who selected 'strongly agree' or 'strongly disagree'. Those who 'strongly disagreed' or 'strongly agreed' with the evolution statement were more likely to choose 'integration' (35 cases in the former; 123 in the latter) than 'dialogue' (13 cases in the former; 74 in the latter). Therefore, denial of or uncertainty towards evolution cannot be 'neatly packaged as scientific knowledge deficit' (Elsdon-Baker, 2015: 425), nor does it necessarily mean that the respondent is anti-science. Moreover, in the case of my sample, it is not clear whether clergy are unsure or doubtful about the *wording* of the statement or about the process of evolution itself. Indeed, clergy could be expressing uncertainty about this evolution statement because there is no mention of God and they thus read it as emphasizing biological evolution at the *expense* of a theistic understanding (on how including references to God in survey items on evolution dramatically alters responses, see Hill, Chapter 2, this volume).

Another question in the ECLAS survey that is relevant here is the following statement-style question: 'Genesis 1 is a scientific account of how God made the world in 6 days' (see Table 4.4). Based on the responses from this question, one might be tempted to suggest that those who ticked 'agree' or 'strongly agree' were more likely to endorse a creationist position. After all, creationist organizations seek to prove *scientifically* that evolution is 'pernicious' and 'incorrect', and use the 'authenticating symbols' of science to support their arguments (Harrold et al, 2004: 73). However, it is important to bear in mind how this question might be interpreted. For example, one respondent ticked 'agree' and wrote next to the question, 'In the sense of it being a scientific model of its time, to be read as "why" rather than "how"'. Another consideration I had to take into account when using this question was whether those ticking 'agree' or 'strongly agree' were being influenced by the question order. Figure 4.1 shows the Barbour typology and asks respondents to indicate how they think science and Christianity should relate to one another (this question came before the Genesis 1 statement). If a respondent ticks 'integration' and elaborates on their choice with a sentence or two (which is what they are required to do), then it is far from inconceivable that they reach the question displayed in Table 4.4 and instinctively tick 'agree' or 'strongly agree' as this would exemplify an integrative approach.

Recent findings, emerging from the 'Science and Religion: Exploring the Spectrum' (SRES) project led by Elsdon-Baker and her team, have pointed towards some significant results. The first is that there are relatively low numbers of religious or spiritual believers who take the creationist stance in the UK (at 9%). This compares to 62% of religious and spiritual believers who accept evolution. Perhaps most surprising, however, was the finding that one in five British atheists express doubts about evolutionary theory's ability to account for human consciousness (Elsdon-Baker et al, 2017). This suggests that those endorsing anti-evolutionary stances are much more diverse than previously thought. One possible reason for this is, in Elsdon-Baker's words, the 'post-truth' zeitgeist, which entails a distrust towards science and evidence-based assessments made about the world. However, as Elsdon-Baker and her colleagues point out, along with sociologists

**Table 4.4: Cross-tabulation for Genesis 1 as a scientific account and biological evolution as the best model for understanding the natural world**

| | | | Biological evolution is the best model we have for understanding the natural world | | | | | |
|---|---|---|---|---|---|---|---|---|
| | | | Strongly disagree | Disagree | Neither agree nor disagree | Agree | Strongly agree | Total |
| Genesis 1 is a scientific account of how God made the world in 6 days | Strongly disagree | Count | 22 | 51 | 108 | 277 | 168 | 626 |
| | | % within question | 3.5 | 8.1 | 17.3 | 44.2 | 26.8 | 100 |
| | Disagree | Count | 9 | 55 | 114 | 121 | 26 | 325 |
| | | % within question | 2.8 | 16.9 | 35.1 | 37.2 | 8.0 | 100 |
| | Neither agree nor disagree | Count | 7 | 8 | 40 | 26 | 6 | 87 |
| | | % within question | 8.0 | 9.2 | 46.0 | 29.9 | 6.9 | 100 |
| | Agree | Count | 6 | 11 | 9 | 8 | 4 | 38 |
| | | % within question | 15.8 | 28.9 | 23.7 | 21.1 | 10.5 | 100 |
| | Strongly agree | Count | 8 | 3 | 5 | 5 | 3 | 24 |
| | | % within question | 33.3 | 12.5 | 20.8 | 20.8 | 12.5 | 100 |
| Total | | Count | 52 | 128 | 276 | 437 | 207 | 1100 |
| | | % within question | 4.7 | 11.6 | 25.1 | 39.7 | 18.8 | 100 |

Evans (2011) and Hill (2014), just because an individual is sceptical about evolution, this does not mean that they extend this scepticism to *all* aspects of science; this is, in fact, extremely rare (Elsdon-Baker, 2017).

Given that the categories of science and religion are porous, it is interesting to look at the cross-tabulation in Table 4.4.[3] The first striking feature of this table is that most clergy surveyed are pro-evolution and do not see Genesis 1 as a scientific account: 592 respondents (53.8%) 'agree' or 'strongly agree' with the biological evolution statement while 'disagreeing' or 'strongly disagreeing' with the Genesis 1 statement, and overall, 951 (86.4%) 'disagree' or 'strongly disagree' that Genesis offers a scientific account of creation. As referenced earlier in the chapter, there is a sizeable group of clergy expressing hesitancy towards evolution but this was found not to be the case in the Genesis 1 statement (25.1% in the former; 7.9% in the latter). There are much smaller numbers of those who might fit with what one might deem the conventional 'creationist' stances (in terms of being anti-evolution and seeing Genesis 1 as a scientific account): these only comprise 28 cases (2.5%) in total. Elsdon-Baker's research has shown that not all those who are characterized as 'creationists' from survey data necessarily hold that particular world view. Therefore, even the 28 clergy that I referred to may not hold creationist views at all, but were simply expressing disagreement with the phrasing of the question. There is also another interesting group, containing 137 cases, who 'disagree' or 'strongly disagree' with biological evolution as the best way of understanding the natural world and also 'disagree' or 'strongly disagree' with Genesis 1 as a scientific account. This group of clergy could lean towards 'intelligent design' but one has to be cautious in drawing conclusions from Table 4.4 since the numbers are relatively low. Most clergy do *not* take an anti-evolution stance, nor do they view Genesis 1 as a scientific account. Moreover, in a similar vein to Hill in Chapter 2, I would argue that there is an assumption on the part of the researcher that their sample must hold 'logically consistent views'; yet, as both Tables 4.3 and 4.4 show, there is a level of messiness and incoherence in public opinion (Hill, Chapter 2, this volume).

## Conclusion

The aim of this chapter has been to set out some of the methodological challenges of conducting research into science and religion. It is clear that the social-scientific study of science and religion (in the UK at least) is in its infancy and projects such as SRES are vital in mapping out the interdisciplinary landscape for this research area. Work continues to be done by scholars globally, including all of those included in this collection. It becomes apparent from carrying out research in this area that, as researchers, we are continuing to challenge the popular discourse around an overarching conflict between science and religion – perpetuated by both media mouthpieces (non-religious and Christian alike) and the social sciences. The research presented in this chapter has shown that science and religion ought to be treated in an exploratory way and that we must be aware of how our own tools of analysis can perpetuate or challenge existing stereotypes about science and religion. Typologies, while useful in terms of creating a framework for how people might understand science and religion, should also be treated with caution: my own research with clergy in England has shown the often complex, multilayered and contradictory approaches that they take to understanding the relationship between science and Christianity.

## Notes

1 The gender balance of the interview participants was also poor, with only two women included in the research. This reflects the current church leadership structures, in which senior roles are typically held by men (the first female bishop was instated in 2014).

2 With a .005 linear-by-linear association.

3 Significant at 0.000 Monte Carlo two-sided test.

## References

Archer, L., DeWitt, J., Osbourne, J., Dillon, J., Willis, B. and Wong, B. (2012) 'Science aspirations, capital, and family habitus: how families shape children's engagement and identification with science', *American Educational Research Journal*, 49(5): 881–908.

Baker, J. (2012) 'Public perceptions of incompatibility between science and religion', *Public Understanding of Science*, 21(3): 340–53.

Barbour, I. (2000) *When science meets religion: Enemies, strangers or partners?*, London: SPCK.

Berger, P. (1970) *A rumour of angels: Modern society and the rediscovery of the supernatural*, London: Allen Lane.

Bouveng, R. and Wilkinson, D. (2016) 'Going beyond the how and why of science–religion? Senior Christian leaders on science and personal faith', *Science and Christian Belief*, 28(2): 100–15.

Cantor, G. and Kenny, C. (2001) 'Barbour's fourfold way: problems with his taxonomy of science–religion relationships', *Zygon*, 36(4): 765–81.

Colburn, A. and Henriques, L. (2006) 'Clergy views on evolution, creationism, science and religion', *Journal of Research in Science Teaching*, 43(4): 419–42.

Crano, W. and Brewer, M. (2002) *Principles and methods of social research*, Mahwah, NJ: Lawrence Erlbaum Associates.

Dickerson, D., Dawkins, K. and Penick, J. (2008) 'Clergy's views of the relationship between science and religious faith and the implications for science education', *Science and Education*, 17(4): 359–386.

Ecklund, E. (2010) *Science vs. religion: What scientists really think*, New York, NY: Oxford University Press.

Ecklund, E., Sorrell, K. and Park, J. (2011) 'Scientists negotiate boundaries between religion and science', *Journal for the Scientific Study of Religion*, 50(3): 552–69.

Elsdon-Baker, F. (2015) 'Creating creationists: the influence of "issues framing" on our understanding of public perceptions of clash narratives between evolutionary science and belief', *Public Understanding of Science*, 24(4): 422–39.

Elsdon-Baker, F. (2017) 'Questioning evolution is neither science denial nor the preserve of creationists', *The Guardian*, 5 September.

Elsdon-Baker, F., Leicht, C., Mason-Wilkes, W., Preece, E. and Piggott, L. (2017) 'Summary report of preliminary findings for a survey of public perspectives on Evolution and the relationship between evolutionary science and religion'. Available at: https://sciencereligionspectrum.org/wp-content/uploads/2017/09/SRESYouGov-survey-preliminary-findings-5.9.17.pdf (accessed 14 April 2018).

Evans, J. (2011) 'Epistemological and moral conflict between religion and science', *Journal for the Scientific Study of Religion*, 50(4): 707–27.

Evans, J. and Evans, M. (2008) 'Religion and science: beyond the epistemological conflict narrative', *Annual Review of Sociology*, 34: 87–105.

Funk, C. and Alper, B. (2015) 'Perception of conflict between science and religion', Pew Research Center. Available at: www.pewinternet.org/2015/10/22/perception-of-conflict-between-science-and-religion/ (accessed 16 April 2018).

Gilliat-Ray, S. (2005) 'Closed worlds: (not) accessing Deobandi *dar ul-uloom* in Britain', *Fieldwork in Religion*, 1(1): 1–33.

Gould, S. (1997) 'Non-overlapping magisteria', *Natural History*, 106(2): 16–22.

Gregory, D. (2017) 'Do science and Baptists mix?', *Baptist Ministers Journal*, 336: 5–11.

Harrold, F., Eve, R. and Taylor, J. (2004) 'Creationism, American-style: ideology, tactics and rhetoric in a social movement', in S. Coleman and L. Carlin (eds) *The cultures of creationism*, Aldershot: Ashgate, pp 67–85.

Hill, J. (2014) 'Rejecting evolution: the role of religion, education, and social networks', *Journal for the Scientific Study of Religion*, 53(3): 575–94.

Hill, J. (2015) 'Do Americans believe science and religion are in conflict?'. Available at: www.bigquestionsonline.com/2015/04/06/americans-believe-science-religion-are-conflict/ (accessed 16 April 2018).

Kelley, P. (2009) 'Stephen Jay Gould's winnowing fork: science, religion, and creationism', in W. Allmon, P. Kelley and R. Ross (eds) *Stephen Jay Gould: Reflections on his life*, Oxford: Oxford University Press, pp 171–89.

Mathieson, K. (2017) 'Rebranding science so it is seen by all people as a fundamental and inclusive part of our society', LSE Impact Blog. Available at: http://blogs.lse.ac.uk/impactofsocialsciences/2017/11/15/rebranding-science-so-it-is-seen-by-all-people-as-a-fundamental-and-inclusive-part-of-our-society/ (accessed 16 April 2018).

McGrath, A. (2016) *Inventing the universe: Why we can't stop talking about science, faith and God*, London: Hodder and Stoughton.

Polkinghorne, J. (1996) *Serious talk: Science and religion in dialogue*, London: S.C.M. Press.

Scheitle, C. (2011) 'US college students' perception of religion and science: conflict, collaboration, or independence? A research note', *Journal for the Scientific Study of Religion*, 50(1): 175–86.

Scheitle, C. and Cornell, N. (2015) 'Hearing clergy speak about social and political issues: examining the effects of religious tradition and personal interest', *Social Science Quarterly*, 96(1): 148–60.

Scheitle, C. and Ecklund, E. (2017) 'Recommending a child enter a STEM career: the role of religion', *Journal of Career Development*, 44(3): 251–6.

Snow, C. (2012) 'The Rede lecture (1959)', in C. Snow and S. Collini (eds) *The two cultures: And a second look*, Cambridge: Cambridge University Press, pp 1–22.

Village, A. and Baker, S. (2013) 'Rejection of Darwinian evolution among churchgoers in England: the effects of psychological type', *Journal for the Social Scientific Study of Religion*, 52(3): 557–72.

PART II

# Belief in the Study of Science and Technology

5

# From 'Science and Religion' to 'Transcendence in Science', or: What We Can Learn from the (History of) Science and Technology Studies

*Silke Gülker*

## Introduction

There is a problem in debates on science and religion: when talking about 'science', on the one hand, and 'religion', on the other, willing or not, one constructs and reconstructs two distinct categories. This is true independent of any positionings towards models of conflict, independence, dialogue, integration (see Barbour, 1997) or non-overlapping magisteria (see Gould, 2011 [1999]), or any other description of this relationship. Simply upon using those categories, we explicitly or implicitly assume that we already know what 'science' means and how it is different from what is not science in general, and different from religion in particular.

This is not to say that there are no such differences. However, both spheres – science and religion – are not as clear-cut and uncontested as the use of such categories might imply. Indeed, we can observe intense and ongoing debates in both sub-disciplines dealing with those categories: the sociology of religion and the sociology of science (or science and technology studies [STS]).

Debates on science and religion seem to be blind, or at least not very attentive, to those ongoing debates. In the case of the sociology of religion, there are a number of authors who doubt that it is even possible to give a definition of what 'religion' is.[1] Others criticize its usage for normative reasons because this term would seem to reproduce dominant power relations between the so-called Western and Eastern worlds (Dubuisson, 2007; Fitzgerald, 2010). In the case of STS, the field was, indeed, founded to call into question the boundary between science and non-science. Thus, while authors in the specialized sociological sub-disciplines put all their effort into deconstructing and criticizing the categories of science and religion, those same categories are often dealt with as given and uncontested when it comes to the dichotomy of science and religion.

This volume, from different angles, seeks to enrich the debate by also questioning the categories. Within this framework, this chapter focuses on one side of this dichotomy, namely, the science side. The underlying question that I address is to what extent 'doing science' might have similarities with 'doing religion'. With this perspective, I look beyond the categories and turn the relationship between science and religion into an empirical – instead of a theoretical or even ideological – question. The very idea of looking empirically at different practices of knowledge-making is relevant, then, not only regarding science and institutionalized religion, but also regarding any form of ultimate belief system.

To approach this question, the chapter looks at the discipline that is specialized in analysing 'doing science': STS. What is striking when surveying contemporary and recent work in this field, however, is how little discussion we find on the role of belief and religion in science – be it religion as organization, institution or practice (see the second section). This is striking because this relationship was one of the most important topics in the early stages of the sociology of science, which is one of the forerunner fields of STS. The chapter discusses the work of two pioneering authors in the sociology of science, Robert K. Merton and Ludwik Fleck. While Merton's work can still be inspiring for contemporary investigations of the relationship between science and religion on a meso- and macro-level, Fleck's concept of 'thought collectives' and 'thought styles' asks for comparative empirical studies on a micro-level (see the third section). Against this background, the fourth section presents an idea of how to implement such micro-level empirical work beyond the science versus religion dichotomy, specifically, by analysing transcendence constructions, demonstrated here in the field of stem cell research. The fifth section summarizes the main points and draws conclusions from the chapter.

## Aim and current work in STS

The aim of STS is to emphasize and analyse the social embeddedness of science.[2] STS is an interdisciplinary field by mission since it was founded to combine different perspectives on science that used to be developed separately in different disciplines, such as the philosophy, history and sociology of science, or the philosophy, history and sociology of technology. Authors in the STS field are convinced that science can be studied neither as a solely epistemic enterprise nor as a solely social process, but that both are intertwined.

Today, we can identify two different strands in the STS field: one focuses mainly on a meso- or partly macro-level, and one focuses mainly on a micro-level. On the meso- and macro-level, STS researchers investigate how either scientific organizations or science as a social system is interrelated with its environment.[3] In recent years, these works have concentrated on the relationship between science and politics (Jasanoff, 1990; Weingart, 2001), science and economy (Etzkowitz et al, 1998; Slaughter and Leslie, 1999), or science and mass media and the broader public (Nowotny et al, 2013; Weingart and Taubert, 2017). A particular environment of political regulations, economical principles and cultural values co-evolve with the evolution of science – this would be the very short summary of a huge body of international literature.

On a micro-level, authors investigate the practice of scientific knowledge-making. In an anthropologist's manner, they go where the work is done, which means in the research laboratory – at least in the natural sciences, where most of the work has its focus. Bruno Latour can be seen as a pioneer of the field; with his early laboratory studies, he founded a new sub-discipline (Latour and Woolgar, 1986).

The main point of these studies is to demonstrate how scientific knowledge is socially constructed through the process of knowledge-making. Karin Knorr-Cetina (1981), for example, describes the many decisions that are taken in order to 'laboratorize' a particular thing that is to be investigated: just by taking objects into the laboratory, scientists change the object, focus on some aspects and leave out others, change the environment, and thereby themselves construct and produce what is to be investigated.

The anthropological strand of the STS field has developed in quite lively and diverse directions, and authors stress different aspects in deconstructing scientific knowledge. Feminist authors demonstrate the role of gender in this process (eg Haraway, 1988; Keller and Longino, 1996), while others focus on the role of class. Today, a growing number of authors with a postcolonial perspective also question the role of race and racism in the process of scientific knowledge-making (Verran, 2010; Schramm, 2017). Thus, as a whole STS is engaged in clarifying the interrelation between science and society, and with demonstrating the role that social constructions in general, and identity constructions in particular, play in the process of scientific knowledge-making.

Now, looking at this field and its programme, it would be more than reasonable to also investigate the role that religion plays in this process of scientific knowledge-making – be it religion as organization, as individual belief or as cultural practice. It seems to be obvious to assume that religious organizations might influence research agendas in many countries. Furthermore, there is no reason to assume that religion, as part of identity construction, has less of an impact on this process than gender or class do. However, we find hardly any examples of STS work that deals with these questions. Looking through the most popular handbooks of the field (eg Jasanoff et al, 1995; Hackett et al, 2008; Felt et al, 2017), we find chapters on many different topics, but we do not find any article investigating the role of belief and religion in science. This does not mean that there are no authors at all working on the mentioned empirical questions.[4] However, the relationship between science and religion is obviously not at the centre of STS's research agenda.

Looking at the ambitious interdisciplinary agenda of STS, this observation is remarkable. It becomes even more striking when we look at the debates that later led to the development of STS. When the sociology of science was founded in the first half of 20th century, the relationship between science and religion was one of the most contested topics. A look back at these early works and debates might give new ideas for contemporary work.

## Forerunner of STS: the sociology of science

### *The early roots of STS: against positivism*

As with any field of investigation, STS did not develop out of nothing, but built on a great deal of work from existing disciplines. The philosophy of science and the sociology of science can be seen as the most important forerunner fields of STS. The founding of the sociology of science, however, can only be understood against the backdrop of intense debates about the relationship between science and religion.

The sociology of science, in its early stage and maybe still today, has one clearly identifiable antagonist, and that is positivism. What is sometimes used as a pejorative term today was a serious and dominating philosophical school in the early 20th century. Although the enthusiasm about Auguste Comte's (1956 [1844]) more ideological version of pure positivism was decreasing, the members of the Vienna Circle developed their idea of 'logical positivism' quite successfully. Their fundamental argument was that all scientific facts need to be rationalized based on empirical observation. In contrast to Comte, they emphasized that there is a need for rationalizing and verification, and that observation, as such, is not enough. However, they stressed the connection between empirical observation and fact. As with Comte's positivism, the Vienna Circle's main statement was also to make a clear distinction between any form of metaphysical belief or religious dogma, on the one hand, and scientific knowledge, on the other.[5]

Later work in the philosophy of science then put into perspective the question of how far anything like a 'pure observation' could even exist. Karl Popper, who criticized the Vienna Circle for their overly empiricist view, was one of the early authors to take into account the scientist's own perspective (Popper, 1959). He stressed that before scientists make any observation, they already refer to a theory. Scientific work, however, remains something that happens solely based on logical work and, in Popper's view, is thereby something that is clearly different from all knowledge-making outside the realm of science.

In the 1930s, two books were published that call into question this idea of the exclusiveness of science in different ways: Robert K. Merton did so on a meso-level and Ludwik Fleck did so on a micro-level. Both authors deal intensively with the relationship between science and religion, as will be demonstrated in the following sections.

## *Forerunner on a meso-level: Robert K. Merton*

Robert K. Merton (1910–2003) is regarded as one of the founders of the sociology of science. His early work on *Science, technology and society in seventeenth-century England* (Merton, 1972 [1938]) became known as the 'Merton Thesis'. In this book, based on his doctoral dissertation in 1936, he emphasizes the role of Puritanism in the rise of the so-called Scientific Revolution.

The book is, in its first part, an empirical study on the shifts in the vocational interests of intellectual elites. Focusing on the time period from 1600 to 1700, Merton analyses biographical notes on 'individuals who achieved a measurable degree of distinction in British history' (Merton, 1972 [1938]: 367) in the *Dictionary of national biography*. As a result, he finds interest rises in all aspects of science while interest in religion declines.

To explain this shift, Merton neither simply refers to the secularization thesis, nor describes science as the winner in a conflict between science and religion. Rather, he argues that a value system particular to Puritanism also led to the rise of interest in sciences. He explains his research interest as follows:

> Puritanism, evoking and shaping the sentiments which pervaded every phase of human action in this period, was the religious movement which notably incorporated dominant cultural values. As such, it provided a measuring rod for the worth of various social activities. What, then, were its relations to science? Did Puritanism, as so often we are told, involve that sort of fervid fanaticism which brooks nothing but its own religious goals? And, if so, what of the cultural implications of such an attitude? What were the consequences for the new science of the powerful motivations which derived from Puritanism? In short, we are concerned with the complex modes of interaction between a religious ethic and science, not as these appear to apologists of the two camps, but as they occurred in the course of actual social development. (Merton, 1972 [1938]: 415)

Thus, similar to Max Weber's (2013 [1920]) thesis on the close connection between Protestantism and capitalism, Merton assumes this connection between Puritanism, particularly in England, and science. There are two aspects of this Puritan ethos that Merton assumes to be responsible for this connection, both expressed in a quote of Robert Boyle, who, according to Merton, wrote in his last will and testament[6]:

> Wishing them also a happy success in their laudable attempts, to discover the true Nature of the Works of God; and praying that they and all other Searchers into Physical Truths, may Cordially refer their Attainments to the Glory of the Great Author of Nature, and to the Comfort of Mankind. (Boyle, quoted in Merton, 1972 [1938]: 4, fn 1)

On the one hand, science means to praise God by discovering the 'true Nature of the Works of God'. Merton illustrates many examples of scientists at the Royal Academy who felt the need to follow exactly this vocation, to praise God by doing science. He also made clear how the Reformation led to great pressure for scientists to feel that their work was in line with their belief in God because the Reformation emphasizes individual responsibility for one's own salvation.

On the other hand, as Boyle's quote points out, science may also refer to the 'Comfort of Mankind', which means to help to solve practical problems. This request refers to the idea of social welfare, which is an important part of Puritan ethos as well. Scientific progress was seen as a major contribution to the well-being of many, to overcome poverty and towards the domination of nature for the good of mankind.

Merton, in this early – and arguably the first – work in the sociology of science, denies the idea of science looking for cognition and truth independent of any social environment. Rather, he stresses how far scientific work is driven by the value systems of the individual scientists, as well as by the organizations that form an institutional environment of the scientific work. Thus, Merton's research agenda appears very similar to the contemporary STS work on a meso- and macro-level. The difference is, however, that the contemporary work only very rarely focuses on religion as a cultural or institutional environment of scientific work. Today's interest lies, as mentioned, much more in the investigation of relationships between science and politics, science and economics, or science and the public, but not in the relationship between science and systems of ultimate belief.

In a way, Merton had already prepared this shift himself. It is not by accident that he chose a historical case study as a method to demonstrate his point. For him, the close interconnection between religious values and scientific interests is a purely historical matter. At many points, he makes clear how alien this 17th-century world of religiously minded scientists must seem to contemporary modern scientists, for example:

> To a modern, comparatively untouched by religious force, and noting the almost complete separation, if not opposition, between science and religion to-day, the recurrence of these pious phrases is apt to signify merely customary usage, and nothing of deep-rooted motivating convictions. (Merton, 1972 [1938]: 5)

Thus, his general argument was not so much about science and religion in particular as about values, institutions, organizations and science in general. Since the secularization thesis became even more popular in the decades after Merton's book was published than it already was in his time, it is not surprising that later sociology of science focused on other issues rather than on the interconnection between religious values and science. Today, however, as we can observe that religion, on an organizational level, is still one of the very powerful actors in most countries worldwide, there is good reason to also investigate its role in the making of science anew. For this, Merton's work can still be inspiring – even if it was merely a starting point for discussion.

Merton's work nevertheless remains at the organizational level of science: he focuses on the question of what motivates scientists to do what they do, and he then stresses the interconnectedness of research interests and the social environment. Once a particular topic is chosen, even Merton would think of scientific work as an exclusive enterprise, which can operate separately from any non-scientific force – at least, he would pose this idea as a normative ideal, as he expresses in his work on the ethos of science (Merton, 1942). Thus, for Merton, the social environment and individual values are interfering factors that are to be controlled in order to conduct real and good science. Therefore, even more interesting for an STS perspective on the interconnectedness of science and religion is the early work of Ludwik Fleck, who posits that the differences between the two are an empirical, not a theoretical, question.

### *Forerunner of micro-level studies: Fleck and the sociology of knowledge*

Ludwik Fleck was a physician and a biologist, and he published a book that later became famous as a pioneering work in the sociology of science: *Genesis and development of a scientific fact* (Fleck, 1979), first published in 1935 in German as *Entstehung und Entwicklung einer wissenschaftlichen Tatsache* (Fleck, 2015 [1935]). Mentioning some aspects of Fleck's biography when talking about this book is helpful for understanding the originality of the book, on the one hand, and its very late reception, on the other. Fleck is probably the first author to emphasize the social process of knowledge-making. He neither wrote as a philosopher who reflected on logical questions, nor investigated the organizational structure from distance. Rather, Fleck reflected on his own work and experience as a biologist inside a research laboratory. In doing so, he was also probably the first author to focus on biological research instead of physics or other natural sciences in order to develop an epistemological contribution. Thus, it was his being a biologist that led to the originality of his work.[7]

The book, original and radical as it was, did not become popular for many years. The reason for this again has to do with his biography, or, better to say, with the political regime of his time. Fleck was a Polish Jew and lived in Lvov (today, Lviv in the Ukraine) when the German National Socialists invaded the country in 1939. He and his family were arrested in Lvov, and he was later forced to continue his work in immunology in concentration camps in Auschwitz and Buchenwald. Fleck survived the Second World War and took up his work in biology again. However, his book in the philosophy/sociology of science was forgotten for many years until Thomas Kuhn mentioned it in the preface of his famous book *The structure of scientific revolutions* in 1962 (Kuhn, 1962).

The very interesting point now is that when the book was rediscovered in the late 1960s and early 1970s, the references that Fleck made regarding the relationship between science and religion were pretty much overlooked. This is probably due to the fact that, at that time, religion did not play an important role as a topic in social science in general, and in the sociology of science in particular.

Which references to the relationship between science and religion do we find in Fleck's book? The main argument of the book is that any cognition is collective in character. Against positivism and the Vienna Circle's philosophy of science, Fleck calls into question the very idea of any individual cognition, as he points out: 'Cognition is therefore not an individual process of any theoretical "particular consciousness". Rather it is the result of a social activity, since the existing stock of knowledge exceeds the range available to any one individual' (Fleck, 1979: 38). For an individual to recognize anything, he or she always needs to build on an already-existing stock of knowledge. A community that shares a particular stock of knowledge and a particular *thought style* is named by Fleck a *thought collective*:

> *If we define 'thought collective' as* a community of persons mutually exchanging ideas of maintaining intellectual interaction, we will find by implication that it also provides the special 'carrier' for the historical development of any field of thought, as for the given stock of knowledge and level of culture. This we have designated thought style. *The thought collective thus supplies the missing component.* (Fleck, 1979: 38, emphases in original)

With these ideas, Fleck argues very early for many points that we will find much later in the work of not only Thomas Kuhn, but also Bruno Latour and in the whole strand of anthropological work in STS. According to Fleck, a stock of knowledge, or cognition, is primarily driven not by a given fact in nature, but rather by the interaction of persons communicating according to a particular thought style. In order to understand the development of science, therefore, one needs to empirically analyse these interactions and to deconstruct the applied thought style.

However, Fleck, in contrast to later work in the STS community, explicitly directs his argument against the idea that there is a substantial difference between scientific or religious recognition. This is not to say that both would be the same to him. Rather, he argues for this radical turn to empirical and comparative investigation – the question for him of what the difference is between scientific and religious recognition, then, is an empirical and not a logical or fundamentally theoretical question.

The characteristic of a thought collective is to produce binding or compulsive truth claims. Fleck describes the role it has for an individual, referring to Durkheim's (2013 [1933]: 31ff) description of collective consciousness. The thought style is not available to an individual; it is just there and cannot be questioned. This, Fleck argues, is true of any thought style and, thus, also of scientific reasoning. The community in general, as well the scientific community, has this power to create dogma-like knowledge. Through collective reasoning, it can produce words that become magic, have a magical power, as Fleck (1979: 43) puts it:

> This social character inherent in the very nature of scientific activity is not without its substantive consequences. Words which formerly were simple terms become slogans: sentences which once were simple statements become calls to battle. This completely alters their socio-cogitative value. They no longer influence the mind through logical meaning – indeed they often act against it – but rather they acquire a magical power and exert a mental influence simply by being used.

According to Fleck, science is a social process, as is any form of knowledge-making. Science, from this point of view, is *one* thought style of *one* collective among others. To find the differences between different styles, be it in science, religion or other spheres, empirical work is needed. This work, however, is not as straightforward as one might assume. Thought styles always imply something implicit, unconscious or transcendent to the individual. Fleck (1979: 104) illustrates this aspect with reference to the initiation into a thought style:

> There is an apprenticeship period for every trade, every religious community, every field of knowledge, during which purely authoritarian suggestion of ideas takes place, irreplaceable by a 'generally rational' organization of ideas. The optimum system of a science, the ultimate organization of its principles, is completely incomprehensible to the novice. Yet this is the only valid yardstick for the expert.... The initiation into any thought style, which also includes the introduction to science, is epistemologically analogous to the initiations we know from ethnology and the history of civilization. Their effect is not merely formal. The Holy Ghost as it were descends upon the novice, who will now be able to see what has hitherto been invisible to him. Such is the result of the assimilation of a thought style.

Science is interaction within a social collective, and to understand how scientific progress works, empirical work is needed to reconstruct this interaction. However, there is a good part in this that remains invisible, both to the interacting subjects and probably, to a degree, for the social scientist observing the interaction as well.

Furthermore, thought styles – independent of where they appear – have the power to construct dogma-like truth:

> At a certain stage of development, the habits and standards of thought will be felt to be the natural and the only possible ones. No further thinking is even possible. But once they have entered personal consciousness, they can also be regarded as supernatural, a dogma, a system of axioms, or even a useful convention. In this context it would be of interest to compare the history of science or the history of sports from semireligious practices in antiquity to the health-oriented sports of our own day. (Fleck, 1979: 107)

Here, Fleck again argues for comparative empirical work to be done in order to illuminate different thought styles at different times and places. Illuminating thought styles, then, means to deconstruct the making of dogmas. This, to conclude Flecks argument, would be of value in science and in religion since there is no fundamental logical difference between the two, but there are empirical differences that have yet to be investigated. Furthermore, the quote emphasizes that, in Fleck's mind, producing dogmas is not an exclusive power of the science system or of institutionalized religions, but a power of any religious and non-religious belief system.

Looking at the work in STS of the last decades, we can notice that Fleck's call for empirical investigation was widely heard and followed. The laboratory studies have exactly this aim: to reconstruct interaction within the laboratory. With her focus on different 'epistemic styles', Knorr-Cetina (1999), for example, also implements the idea of comparing different modes of interaction, or, in Fleck's terms, different thought styles. We furthermore find empirical analyses on different disciplines, such as the social sciences (Lamont, 2005) or mathematics (Heintz, 2000). However, the anthropological work in STS remains within the realm of science, comparing different disciplines regarding their modes of knowledge-making. There is obviously little interest in the very idea of comparing thought styles in science, religion or other thought collectives based on particular belief systems.

Again, the argument of this chapter is not that there is no difference between science and religion. Rather, it stresses that the boundaries between the two are not clear-cut, but, indeed, always contested. Talking theoretically about science, on the one hand, and religion, on the other, will, in the end and in Fleck's terms, reproduce dogmas of the particular 'science and religion thought collective'.

Against this background, in the last part of this chapter, I will present some ideas on how to implement the notion of empirically analysing different thought styles. I will do so by referring to a core concept in the sociology of knowledge of Alfred Schütz and Thomas Luckmann: the construction of transcendence.

## How the debate can get enriched: transcendence in science as a research perspective

Fleck, with his early pioneering work, calls for empirical work to better understand how different thought styles function – whether in science, religion or other fields. As outlined earlier, a lot has been accomplished within the realm of science over the last decades. However, those investigations do not ask questions regarding the relationship between science and religion.

In the following section, I will suggest a research perspective that also does not put the relationship between science and religion as such in the centre. Instead, it focuses on a concept that is important in both science and religion: the construction of transcendence. In doing so, it seeks to overcome the dichotomy of two categories and focus on the fundamentally important question of how boundaries are constructed between something that is *available*[8] and something that is *not available*. The construction of such boundaries is important both in the realm of religion and, as I will demonstrate, in science.

I will, very briefly, present a study that inquires into the role of transcendence constructions in stem cell research. Coming from a different angle, namely, the phenomenologically oriented sociology of knowledge, this perspective follows up on Fleck's idea of reconstructing the making of dogmas in a thought style. As described, there is always something about a thought style that remains unavailable to the member of a thought collective, which means here to a scientist in a particular environment. This unavailability is partly due to ideas and contents – Fleck emphasizes the making of unquestionable dogmas within a thought collective. Furthermore, there is an unavailability due to emotions, implicitly and unconsciously taken-for-granted ways of thinking. For a social scientist, it is difficult to reconstruct this making of dogmas; he or she tries to observe something that remains unconscious to the observed subjects themselves.

The idea is that an analysis of transcendence constructions, oriented towards the concept of small, intermediate and great transcendences, can be a way to get closer to these implicit processes. A few words about this concept are needed first, before I demonstrate with an empirical application.

## *Transcendence constructions: the concept*

In normal parlance, 'transcendence' means anything that is other-worldly. The term derives from the Latin *transcendere*, meaning 'to go beyond' any form of limit. However, today, a narrower meaning has become widespread whereby the limit to be exceeded is of a fundamental nature and leads to the hereafter. By contrasting transcendence to immanence, there is an implication that a clear distinction can be made between the inner-worldly and graspable, on the one side, and what lies outside of this world and perception, on the other. Nowadays, transcendence is often, therefore, used in close connection to religion, with religion being the organization that helps to deal with the transcendence of the hereafter.[9]

In early philosophy, however, the term 'transcendence' always had a double connotation: an epistemological and a theological meaning. It was Immanuel Kant (1968 [1781]) who introduced the clear-cut distinction between the two, naming the first transcenden*tal* and the latter transcendent. The earlier two-sided meaning makes the term so valuable for investigating potential similarities and differences between thought styles in science and religion. The concept of Alfred Schütz and Thomas Luckmann (2017 [1979]), in a very particular way, brings the epistemological meaning of transcendence back into the debate.

In the life-world, every actual experience points to something that – within this actual experience – remains imperceptible. Schütz and Luckmann made this aspect a central point of departure of their sociology of knowledge. In their book *Strukturen der Lebenswelt* (in English, *The structures of the life-world* [Schütz and Luckmann, 1973]), they devote a chapter to the borders of experience, and they use the term 'transcendence' both for inner-worldly unavailabilities and for the other-worldly in the sense mentioned earlier.

Schütz and Luckmann emphasize the processual character of transcendence, and they describe gradual but not substantive differences between what they call 'small', 'intermediate' and 'great' transcendences. In all of these cases, this means that in an ongoing experience, something is referenced – a relationship is created to something that is not available in the experience taking place at that moment.

What Schütz and Luckmann describe for the everyday life-world experience can also be presumed for experiences within a research laboratory. An experiment here also points beyond the contemporary, available experience – the available data always need interpretation based on larger concepts that are not part of actual experimentation. For an analysis in stem cell research, the intermediate and great transcendences are particularly important, but in order to comprehend the principle of continuity, it is also necessary to take a brief look at 'small' transcendences. In the case of *small transcendences*, what is currently not perceivable is simply unavailable for reasons of spatio-temporal limits – an experience I can remember could be repeated, and space limits can also, in principle, be overcome.

*Intermediate transcendences* refer to the boundary between me and my counterpart. No matter how well I can imagine it, I will never be able to completely grasp what another person is feeling or thinking, just because of his or her self-being. I will, however, try to interpret what I can see, for example, a facial expression. Based on what I can actually see or otherwise experience, I will try to construct what this might indicate inside the other self-being. However, there will always remain a boundary that I will not be able to cross, something that remains inside the other self-being and outside my experience.

Thus, intermediate transcendences describe what is presumed to be fundamentally unavailable in the counterpart. This unavailability is, however, practically imperceptible in everyday life; what remains unavailable is still demonstrated to a certain extent so that the border is normally hardly perceptible and only becomes clear in crises of communication. This is different in matters concerning *great transcendences*. Schütz and Luckmann call these 'other realities' and describe their characteristics using the example of dreams or trances. Other realities, in this sense, are not available to conscious access, a dream experience cannot be planned, neither is it unequivocally rememberable or repeatable. Such great transcendences, which are experienced in an inner-worldly manner, according to Schütz and Luckmann, can also form the basis for dealing with the ultimate border, namely, the border to our own death. Conceptions about whether anything – and, if so, what – could exist beyond this border can be deduced from experiences of other realities such as dreams and trances.

Now, transcendence, in Schütz and Luckmann's concept, is something given in any experience. The overall aim of their book is to find general anthropological constants, and the experience of boundaries to transcendent spheres is one of these constants. The idea of using this concept to analyse transcendence constructions in stem cell research, however, is based on the assumption that the experiences of these boundaries are also not just given; rather, they are contested and constructed. Science in general, and stem cell research in particular, can be seen as an enterprise in order to make the world more and more available – or to shift the boundary between availability and unavailable further and further. Unavailability thereby always has a double connotation: something can be unavailable or transcendent for technical or practical reasons, and something can be unavailable for ethical reasons. The question, therefore, is how those boundaries are constructed and what remains unavailable even in this dynamic field of making available. The following section illustrates a few examples in this line of thinking.

## *Empirical examples*

The following illustrations are based on two empirical case studies in stem cell research laboratories, one located in the US and one in

Germany. Within each study, a four-month period of ethnographic observation work was combined with biographical interviews. Numerous daily routines of the researchers and experiments were observed, discussed with the scientists carrying them out, audio-recorded and reproduced in daily field notes.[10] The narrative biographical interviews were generally set up according to the concept created by Schütze (1983): with an open starting question, the researchers were asked to tell their life stories. Following the life-story narrative, all interviews included standardized questions on potential ethical conflicts and on the regulation of stem cell research. In total, 38 (US = 28; Germany = 10) scientists participated in the two case studies, coming from 17 different countries.

Thus, the basic methodological approach of the empirical investigation is in line with the laboratory studies in the framework of STS. The main question, however, is not how the researchers 'laboratorize' the world, or how they make it available. Rather, the study is focused on what remains *un*available or transcendent to them, and on how the boundaries between availability and unavailability are constructed and shifted. With this aim, the concept of *intermediate transcendences* is of particular interest because it also brings a new perspective into the debate on the status of non-human entities. There is an ongoing debate on whether and to what degree entities, such as machines, animals or any research objects that take part in the process of investigation, should be recognized as subjects, actors or actants. The different terms refer to different conceptualizations of these entities and their assumed availability to take over action.[11]

The construction of intermediate transcendence, meaning to assume an inner self-being in my counterpart that, at least in part, remains unavailable to me, is the basic operation for including an entity in the social world (*Sozialwelt*) (see Lindemann, 2009). In the laboratory studies, I analysed the degree to which the researchers construct such intermediate transcendences regarding laboratory animals, on the one hand, and cells, on the other.

In short, I found out that there is a difference between the two entities: animals, within a research laboratory, somehow need to be seen as material, as something available. This is due to the logic of experiments: the animals are the container in which a cellular process is studied. However, the observation shows that this construction of animals as available comes up against clear ethical boundaries. Whenever animals give signals of life, the researchers handle them as self-being counterparts and thereby construct intermediate transcendence.

Cells, in contrast, are those entities that are to be discovered in a research laboratory. Thus, all work in the laboratory is about making them available, but their unavailability is what drives the research. However, cells are often constructed as unavailable in principle: researchers talk about them emotionally, and they talk about the cells' own mind and how they are fascinated by them. A big part of the fascination is that those cells are seen as the entities that have the potentiality to change the world.

Thus, cells also incorporate a connection between intermediate and *great transcendence* as I operationalize it in my study. In this context, great transcendence means something that is indicated inside the research laboratory but nevertheless remains unavailable as a whole because it lies beyond the border to the world outside the laboratory. The question in relation to great transcendence is thereby how much the researchers construct the world outside the laboratory as available or unavailable through their scientific work.

Just to give one example to illustrate this idea, the very aim of stem cell research is to find a cure for diseases, or, in other words, to make diseases available. As I find in my study, however, researchers construct diseases very differently according to the particular projects that they are working on. Based on what they can see and measure, researchers try to model a particular part of the disease inside the laboratory. The disease as a whole, however, remains something unavailable outside the laboratory. Some researchers even describe ethical conflicts regarding the making available of diseases in the long run: in their views, science should not try to find cures for all diseases because nature 'knows better' how to regulate life and death.

These are only very few and brief insights into the empirical findings of a larger research project. The aim here is to demonstrate that inside a research laboratory, where it is all about making things available, constructions of unavailability and transcendence also play a significant role. Those constructions are unquestionable for the researchers; they remain unconscious and self-evident and thereby are part of what Fleck describes as a thought style.

The construction of transcendence is a basic operation, important both in the religious and in the scientific spheres. Empirical analysis of those constructions will thus help to overcome a dichotomy of two categories and thereby shed a new light on the relationship between science and religion from within the making of science, religion or any system of ultimate belief.

## Summary and outlook

The departure point for this chapter was that contemporary debates on the relationship between science and religion often risk implicitly or explicitly reproducing two dichotomous categories – even though the content of these categories is highly disputed in the sociology of religion and in STS. The chapter therefore looks for ways of how to overcome this dichotomization and asks how far the concepts of STS can enrich the debate. Authors in this field emphasize the social embeddedness of science and thereby question a clear boundary between science and non-science. Even though there is little work on the role of belief and religion in contemporary STS work, it was one of the main topics when the sociology of science was founded in the early 20th century. Robert K. Merton, who focuses mainly on a meso-level, and Ludwik Fleck, who focuses on a micro-level, can be seen as pioneers of STS, and both reflected intensely on the relationship between science and religion.

The main point to learn from this review is that the relationship between science and religion, or science and any belief system, should be investigated empirically – instead of assuming clear-cut theoretical differences. In doing so, we will find differences and similarities between the different spheres. The example given in this article focuses on a micro-level and illustrates how stem cell research is oriented to making the world more and more available. At the same time, in stem cell research, the construction of unavailability and transcendence also plays a role.

In this context, stem cell research is of particular interest because of its scientific as well as ethical importance. However, many more important fields and topics are to be addressed empirically. The role of belief in economy would be one important example; its role in sociology would be another. Most promising, then, would be a perspective that is able to combine Fleck's micro- and Merton's macro-level focus. Obviously, both are intertwined: the making of dogmas – of concepts that are unquestionable for a given collective – and the construction of boundaries between what is seen as available and unavailable result from complex, dynamic and interwoven processes of individual positionings, interactions on a micro-level and institutionalizations on a meso- and macro-level. The sociological debate on science and religion and belief would be enriched from studies trying to empirically investigate and disentangle these processes.

## Notes

1 For the history of the term 'religion', see Wagner (1991). For a discussion on whether or not sociology needs a definition, see Pollack (1995).

2 For introductions to STS, see Sismondo (2011) and Beck et al (2014).

3 In Germany, this strand has long been dominated by scholars in the tradition of Niklas Luhmann's system theory (see Luhmann, 1992).

4 For example, in medical anthropology, there is a lot of work on the role of belief in the field of medicine (see, eg, Cobb et al, 2012; Peteet and Balboni, 2017). An interesting case is Latour's (2014) book on religious speech, which is about differences and similarities in science and religion. The STS community seems to have some trouble integrating this work into the field (see Golinski, 2010).

5 For an introduction into the philosophy of positivism, see Hacking (1983: 41ff). On the Vienna Circle, see Stadler (2001).

6 Quoted by Gilbert, Lord Bishop of Sarum, *A sermon preached at the funeral of the hon. Robert Boyle* (London, 1692), p 25.

7 For details on Fleck's biography, see the Introduction to the German edition of Fleck's book by Schäfer and Schnelle (2015 [1935]).

8 The German translation of the terms 'available/unavailable' would be '*verfügbar/unverfügbar*', which is a stronger expression than the English term. *Unverfügbar* describes something that, in principle, not only for technical, but also for ethical or even cosmological, reasons, remains outside the realm of human intervention.

9 For an introduction to the history of the term 'transcendence', see Halfwassen (2004), Enders (2004) and Lambrecht (2012).

10 Thus, the work is in line with studies in ethnomethodology (see, eg, Garfinkel et al, 1981; Lynch, 1985).

11 Most famous in this respect became the actor-network-theory of Latour et al (see Law and Hassard, 1999; Latour, 2005); on the debate on machines and action, see also Rammert and Schulz-Schaeffer, 2004; Schulz-Schaeffer, 2007).

## References

Barbour, I.G. (1997) *Religion and science*, New York, NY: HarperCollins.

Beck, S., Niewöhner, J. and Sörensen, E. (eds) (2014) *Science and technology studies: Eine sozialanthropologische einführung*, Bielefeld: Transcript Verlag.

Cobb, M.R., Puchalski, C.M. and Rumbold, B.D. (eds) (2012) *Oxford textbook of spirituality in healthcare*, Oxford: Oxford University Press.

Comte, A. (1956 [1844]) *Rede über den geist des positivismus*, Hamburg: Felix Meiner Verlag.

Dubuisson, D. (2007) *The Western construction of religion: Myths, knowledge and ideology*, Baltimore, MD: John Hopkins University Press.

Durkheim, E. (2013 [1933]) *The division of labour in society*, Basingstoke: Palgrave Macmillan.

Enders, M. (2004) 'Transzendenz: Neuzeit', in J. Ritter, K. Gründer and G. Gottfried (eds) *Historisches wörterbuch der philosophie*, Basel: Schwabe, pp 1447–55.

Etzkowitz, H., Webster, A. and Healey, P. (eds) (1998) *Capitalizing knowledge: New intersections of industry and academia*, New York, NY: State University of New York.

Felt, U., Fouché, R., Miller, C.A. and Smith-Doerr, L. (eds) (2017) *The handbook of science and technology studies* (4th edn), Cambridge, MA: MIT Press.

Fitzgerald, T. (2010) *Discourse on civility and barbarity: A critical history of religion and related categories*, Oxford: Oxford University Press.

Fleck, L. (1979) *Genesis and development of a scientific fact*, Chicago, IL: University of Chicago Press.

Fleck, L. (2015 [1935]) *Entstehung und entwicklung einer wissenschaftlichen tatsache: Einführung in die lehre vom denkstil und denkkollektiv*, Frankfurt am Main: Suhrkamp.

Garfinkel, H., Lynch, M. and Livingston, E. (1981) 'The work of a discovering science construed with materials from the optically discovered pulsar', *Philosophie of Social Science*, 11(2):131–58.

Golinski, J. (2010) 'Science and religion in postmodern perspective: the case of Bruno Latour', in T. Dixon, G. Cantor and S. Pumfrey (eds) *Science and religion: New historical perspectives*, Cambridge: Cambridge University Press, pp 50–68.

Gould, S.J. (2011 [1999]) *Rocks of ages: Science and religion in the fullness of life*, New York, NY: Ballantine Books.

Hackett, E.J., Amsterdamska, O., Lynch, M. and Wajcman, J. (eds) (2008) *The handbook of science and technology studies* (3rd edn), Cambridge, MA: MIT Press.

Hacking, I. (1983) *Representing and intervening: Introductory topics in the philosophy of natural science*, Cambridge: Cambridge University Press.

Halfwassen, J. (2004) 'Transzendenz', in J. Ritter, K. Gründer and G. Gottfried (eds) *Historisches wörterbuch der philosophie*, Basel: Schwabe, pp 1442–7.

Haraway, D.J. (1988) 'Situated knowledges: the science question in feminism and the privilege of partial perspective', *Feminist Studies*, 14(3): 575–99.

Heintz, B. (2000) *Die innenwelt der mathematik: Zur kultur und praxis einer beweisenden disziplin*, Wien: Springer.

Jasanoff, S. (1990) *The fifth branch: Science advisors as policymakers*, Cambridge, MA: Harvard University Press.

Jasanoff, S., Markle, G.E., Petersen, J.C. and Pinch, T. (eds) (1995) *Handbook of science and technology studies* (rev edn), Thousand Oaks, CA: SAGE Publications.

Kant, I. (1968 [1781]) 'Kritik der reinen Vernunft: 2 Auflage 1787', in Preußischen Akademie der Wissenschaften (ed) *Kants werke, akademie ausgabe: Unveränderter photomechanischer abdruck des textes der von der preußischen akademie der wissenschaften 1902 begonnenen ausgabe von kants gesammelten schriften*, Berlin: Walter de Gruyter and Co.

Keller, E.F. and Longino, H.E. (1996) *Feminism and science*, Oxford: Oxford University Press.

Knorr-Cetina, K. (1981) *The manufacture of knowledge*, Oxford: Pergamon Press.

Knorr-Cetina, K. (1999) *Epistemic cultures: How sciences makes knowledge*, Cambridge, MA: Harvard University Press.

Kuhn, T.S. (1962) *The structure of scientific revolutions*, Chicago, IL: University of Chicago Press.

Lambrecht, J. (2012) *Transzendenz: Eine systemanalytische studie*, Würzburg: Königshausen and Neumann.

Lamont, M. (2005) 'Peer evaluation in the social sciences and the humanities compared: the United States, the United Kingdom and France', report prepared for the Social Sciences and Humanities Research Council of Canada. Available at: http://citeseerx.ist.psu.edu/viewdoc/download?doi=10.1.1.529.4425&rep=rep1&type=pdf

Latour, B. (2005) *Reassembling the social: An introduction to actor-network-theory*, Oxford: Oxford University Press.

Latour, B. (2014) *Rejoicing, or the torments of religious speech*, Cambridge: Polity.

Latour, B. and Woolgar, S. (1986) *Laboratory life: The construction of scientific facts* (2nd edn), Princeton, NJ: Princeton University Press.

Law, J. and Hassard, J. (1999) *Actor network theory and after*, Boston, MA: Blackwell Publishers.

Lindemann, G. (2009) *Das soziale von seinen grenzen her denke: 1 aufl*, Weilerswist: Velbrück Wiss (Velbrück Wissenschaft).

Luhmann, N. (1992) *Die wissenschaft der gesellschaft*, Frankfurt am Main: Suhrkamp.

Lynch, M. (1985) *Art and artifact in laboratory science: A study of shop work and shop talk in a research laboratory*, London and Boston, MA: Routledge and Kegan Paul.

Merton, R.K. (1942) 'The normative structure of science', in R.K. Merton and N.W. Storer (eds) *The sociology of science: Theoretical and empirical investigations*, Chicago, IL: University of Chicago Press, pp 267–78.

Merton, R.K. (1972 [1938]) *Science, technology and society in seventeenth-century England*, New York, NY: Fertig.

Nowotny, H., Scott, P. and Gibbons, M. (2013) *Re-thinking science: Knowledge and the public in an age of uncertainty*, Hoboken: Wiley.

Peteet, J.R. and Balboni, M.J. (eds) (2017) *Spirituality and religion within the culture of medicine: From evidence to practice*, New York, NY: Oxford University Press.

Pollack, D. (1995) 'Was ist Religion?', *Zeitschrift für Religionswissenschaft*, 3(2): 163–90.

Popper, K.R. (1959) *The logic of scientific discovery*, New York, NY: Hutchinson and Co.

Rammert, W. and Schulz-Schaeffer, I. (2004) *Können maschinen handeln? Soziologische beiträge zum verhältnis von mensch und technik*, Frankfurt: Campus.

Schäfer, L. and Schnelle, T. (2015 [1935]) 'Ludwik Flecks Begründung der soziologischen Betrachtungsweise in der Wissenschaftstheorie', in L. Schafer and T. Schnelle (eds) *Ludwik Fleck: Entstehung und Entwicklung einer wissenschaftlichen Tatsache: Einführung in die Lehre vom Denkstil und Denkkollektiv*, Frankfurt am Main: Suhrkamp, pp vii–xlix.

Schramm, K. (2017) 'Postkoloniale STS: Einfuehrung', in S. Bauer, T. Heinemann and T. Lemke (eds) *Science and technology studies: Klassische positionen und aktuelle perspektiven*, Berlin: Suhrkamp, pp 471–94.

Schulz-Schaeffer, I. (2007) *Zugeschriebene handlungen: Ein beitrag zur theorie sozialen handelns*, Weilerswist: Velbrück Wissenschaft.

Schütz, A. and Luckmann, T. (1973) *The structures of the life-world*, Evanston, IL: Northwestern University Press.

Schütz, A. and Luckmann, T. (2017 [1979]) *Strukturen der Lebenswelt*, Konstanz: UVK.

Schütze, F. (1983) 'Biographieforschung und narratives interview', *Neue Praxis*, 13(3): 283–93.

Sismondo, S. (2011) *An introduction to science and technology studies*, New York, NY: John Wiley and Sons.

Slaughter, S. and Leslie, L.L. (1999) *Academic capitalism: Politics, policies, and the entrepreneurial university*, Baltimore, MD, and London: Johns Hopkins University Press.

Stadler, F. (2001) *The Vienna Circle: Studies in the origins, development, and influence of logical empiricism*, Wien: Springer-Verlag Wien.

Verran, H. (2010) *Science and an African logic*, Chicago, IL: University of Chicago Press.

Wagner, F. (1991) *Was ist religion? Studien zu ihrem begriff und thema in geschichte und gegenwart*, Gütersloh: Gerd Mohn.

Weber, M. (2013 [1920]) *The protestant ethic and the spirit of capitalism*, Abington: Royal Asiatic Society.

Weingart, P. (2001) *Die stunde der wahrheit? Zum verhältnis der wissenschaft zu politik, wirtschaft und medien in der wissensgesellschaft*, Weilerswist: Velbrück Wissenschaft.

Weingart, P. and Taubert, N.C. (2017) *The future of scholarly publishing: Open access and the economics of digitization*, Cape Town, South Africa: African Minds.

6

# Rational Believers: Religion, Tradition and Spirituality among Indian Scientists

*Renny Thomas*

## Introduction

From February to December 2012, I worked at a laboratory in the Indian Institute of Science (IISc), Bengaluru – one of the leading scientific research institutions in India – as a lab member/visiting researcher. This enabled me to interview and converse with lab members and scientists from various departments of IISc. I revisited the field in 2016. IISc is the habitat of India's elite scientists; some of them are world-renowned scientists in their field. Most of the scientists who work there obtained their PhDs and post-doctoral training from leading American and British universities. The majority of the IISc scientists are Hindus, with a few Christians. The scientists I had conversations with were mostly Hindus, though I interviewed some Christians as well.

This chapter attempts to understand the various ways in which Indian scientists defined their religious life by claiming a distinct religious identity – distinct, specifically, from laypeople/non-scientists. As we will see, they make this distinction not by saying that they are non-believers or atheists; in fact, they have no inhibition in demonstrating their devotion by participating in various religious festivals and rituals. Rather, the distinction relates to the *manner* of their belief and the practices they follow.

The scientists in my study cautioned against comparing science and religion and did not perceive any conflict between the two domains. For them, both these realms are important in their life, and they claimed not to see science through the eyes of religion, or religion through the eyes of science. Many scientists admitted that religion cannot be explained using the 'scientific' method. I argue that for many of these scientists, the relationship between science and religion is about neither the pursuit to prove the existence of God, nor the search for compatibilities between the two. Rather, they considered such enquiries superfluous and wholly unnecessary. For them, religion cannot be conceptualized from an 'objective' point of view since religion and science occupy different realms. At the same time, they do not separate or posit science and religion in oppositional terms as they think their religious beliefs and practices help them to do better science.

Although these scientists have no inhibition in demonstrating their religiosity, we will see that they do make a clear attempt to distinguish their belief as scientists from that of non-scientists. As we will see, they try to be 'rational believers'. This is the theme that I explore in this chapter: how they claim a distinct identity as scientist-believers, as well as their relationship with Indian tradition and the ways in which they express this. They are very proud of their tradition and culture, and they have no inhibition in saying as much. Before we move to their engagement with 'culture' and 'tradition',[1] though, it is imperative to see how they exhibit their love for God in their scientific writings.[2]

## Doing science, acknowledging God

At IISc, I spoke mainly to established senior scientists. To understand how young scientists engage with the question of religion and God, I decided to do archival work in the library. I was interested in how young scientists acknowledged their love for God when they submitted their doctoral work. I searched for old and recently submitted PhD dissertations at the IISc's annex library. Most theses had some such references, acknowledging God for enabling them to finish their work. Historian of science Ronald L. Numbers talks about the references to God in scientific writings in the West during the early years of the debates over Darwin's theories. He argues that it was not unusual for even scientific participants to occasionally allude to God (Numbers, 2007: 239). By the 1880s, however, references to God seldom appeared in the increasingly specialized literature of science in the West (Numbers, 2007: 240). Numbers (2007: 240) writes that 'In

1885 the science popularizer John Fiske remarked that scientific texts were no proper place to refer to the Divine Architect or the Great Designer'. The absence of the reference of God in scientific writings was due to the process of privatization, or the separation of religion from the profession of science (Numbers, 2007: 238). As Numbers (2007: 240) puts it: 'Advocates of privatization argued that separation would be for the comfort and advantage of both science and religion'.

It is now extremely rare to find references to God in Western scientific research publications. In contrast, we often see renowned Indian scientists using metaphors, quotes and examples from religious texts in their advanced scientific writings. Thanu Padmanabhan, one of India's most influential astrophysicists and cosmologists, starts his influential book on theoretical astrophysics with a quote in Sanskrit from The Bhagavad Gita:

> *Yadhyadh vibuthimatsthwam srimadhurjithameva va*
> *Thaththadevava gachchatwam mama tejoamsa sambhavam*
>
> [Whatever that is glorious, prosperous or powerful anywhere, know that to be a manifestation of a part of My splendour.] (Bhagavad Gita, ch 10, verse 41, quoted in Padmanabhan, 2001: xiii)

Muktha Bharathy submitted her PhD in solid state and structural chemistry at IISc in 2006. Her PhD thesis is titled 'Layered oxides and phosphates of bismuth: new structural types and related properties'. In the first page of the thesis, she writes:

> *Poorna Madhah Poorna Midham, Poornath Poornamidhachathe*
> *Poornasya Poornamadaya, Poornamevavasishyathe!*
>
> [The Brahman is infinite; this universe too is infinite. The infinites (universe) emanates from the infinite (Brahman). Assimilating the infinitude of the infinite (universe), the infinite (Brahman) alone is left!] (Brihadaranyaka Upanishad,[3] quoted in Bharathy, 2006)

Similarly, Susanta Kumar Nayak submitted his PhD in solid state chemistry in 2010 on 'Dosorder-polymorphism and co-crystal formation in molecular crystals: an in-depth study in terms of weak intra and intermolecular interactions'. He writes in the first page of his thesis:

*Guru Brahma Guru Vishnu*
*Guru Devo Maheshwara*
*Guru Sakshat Param Brahma*
*Tasmai Shri Guruve Namah*

[Guru is Brahma, Guru is Vishnu, Guru is Lord Maheshwara, Guru is verily the supreme reality. Sublime Prostrations to Him.] (Nayak, 2010)

This extends beyond Hindu scientists. Reuben T. Chacko submitted his PhD in inorganic and physical chemistry in 2007 on 'Functionalization of hyperbranched polythers'. He writes in the acknowledgement page of his thesis:

> It is the glory of God to conceal things, but the glory of king is to search for things – (Proverbs 25:2, The Bible).... The final word of thanks is reserved for my parents and my sister for their overflowing love, encouragement, patience and understanding and above all to God, who we all lovingly call Abba Father! (Chacko, 2007)

On the first page of her 1997 PhD thesis 'Insertion reactions of copper (I) aryl oxides with heterocumulenes', Christina Bhaskaran also writes:

To the only God our saviour be glory
Majesty, power and authority, through Jesus Christ our Lord, before all ages,
Now and for evermore!
For you are great and do marvelous deeds;
You alone are God
I will praise you, O Lord my God,
With all my heart;
I will glorify your name for ever,
You O Lord, have helped me and comforted me
From Him and through Him and to Him are all things
To Him be the glory forever! Amen. (Bhaskaran, 1997)

Then, on the acknowledgement page, she writes again: 'I owe everything to my Lord Jesus Christ – my faithful Friend. With deep sense of gratitude, I offer this work to him. Lord Jesus, please accept this humble offering' (Bhaskaran, 1997).

In acknowledging their gratitude to God, these young scientists did not shy away from vocally expressing their adulation. Here are some further (translated) excerpts: 'I certainly thank GOD for providing this opportunity to live life in this beautiful campus'; 'I thank the Almighty God for bringing me thus far'; 'My thesis is dedicated to Goddess Madurai Meenakshi and God Palani Murugan'; and 'A final word of thanks to God almighty for being with me throughout'.

These quotes testify to the ease with which these young scientists, in their very first piece of official scientific writing, had been able to express their gratitude to God. It shows that they lack fear of any social sanction in acknowledging God as they enter the profession of science. They know that they will be judged not based on their belief in God, but based on their scientific contributions. When I spoke to a senior professor about his PhD students acknowledging God in dissertations, he told me that it is not a problem if young scientists acknowledge God, as the acknowledgement page has no connection with the scientific ideas discussed and so does not affect them. Although it is not *expected* for a PhD student to acknowledge God in their dissertations, it is accepted, as a culture, to acknowledge God in many Indian scientific institutions, including IISc.

This manner of acknowledgement also reflects these scientists' attempts to fence their inner and spiritual domain from science and scientific practices without impacting either. I argue that scientists impose this separation of domains in order to navigate more easily the world of science while at the same time remaining religious and pious. Yet, while being religious, the scientists also try to maintain a distinct identity from non-scientists and laymen. The maintenance of this distinction enables them to position themselves as experts and assume a privileged position in society. They argue that their belief in God is *rational* and hence different from that of other devotees. It is this sense of distinction of belief among scientists that I focus on in the following section.

## Scientists as rational believers

Scientist interviewees, in order to separate themselves from non-scientists, defined religious and spiritual life in their own way to make it more 'rational'. Anthropologist Stefen Helmreich (1997: 369), in his study of American computer scientists and biologists, shows that the scientists find 'Eastern religions' more appealing than Christianity or Judaism and 'often compared their scientific appreciation of the natural world to perspectives informed by "Eastern" religions'. Similarly,

Richard Cimino (2014) has shown how immigrant Muslim, Hindu and Sikh engineers in the US dealt with religion distinctly as applied scientists and engineers, emphasizing the 'rationality' and 'scientific nature' of the belief systems that they followed.

During my interaction with scientists, I observed that they made a distinction between their idea of religion and their view of the non-scientists' practice of religion. They stated that they are not like the non-scientists because scientists are more *rational* in their act of believing. Even when they stated that they believed in God and practised various rituals, they made a point to delineate their differences from others by stressing the spiritual over the ritual and criticizing the superstitious beliefs of the 'lowly folk', which paled in comparison to their more rational and objective belief in God. Robert Bellah (1971: 40) argued that intellectuals are sceptical of the religion that is current among the common people. As he succinctly put it, 'The intellectual has always needed to find a religious form which genuinely expressed his own individuality' (Bellah, 1971: 42). This perceived distinction in belief and practice is thus a marker of these scientists' cultural and academic capital. The distinction that they make has to do with their professional status as elite scientists: as established scientists, they have to be different from others, whether through lifestyle or religious life.

Geneticist Siddhartha[4] is a deeply religious scientist who does *puja*[5] and visits temple every week. However, he, like many other scientists, argued that he is different from the layman in the way he thinks and practises religion. He said:

> "There is a difference between scientists and the general public. For example, when I go to temple and pilgrimage centres, what I see is a rush to go near the sanctum sanctorum. People want to be there physically. But I don't do that. I don't think it is required. For example, in Tirupati temple, most of the time, I skip the line; I don't go to see the sanctum sanctorum. I always stay outside and pray. It is not important to go in the line and see the sanctum sanctorum. In fact, what is important is to go to the temple premises, not necessary that one should go inside and pray in front of the sanctum sanctorum. I pray from outside the temple and come back. I don't think that I am less blessed than those who go inside the temple."

For him, prayer is not something that needs to be performed in front of the statues and images of gods and deities. Prayer is a voluntary act, which need not follow the established dogmatic rules. However, he believed in the act of prayer and thought that it is important to pray to get wishes fulfilled, even if the way he chose to pray differed. The act of distinction is again clear here. He goes to temple, but unlike a lay believer, he does not like to be in the queue to have a look at the deity and pray. As he said: "[W]hat is important is to go to the temple premises, not necessary that one should go inside and pray in front of the sanctum sanctorum". As Marcel Mauss (2003: 54) observed in a different context:

> Every prayer is an act. It is not just dreaming about a myth or simple speculation on dogma, but it always implies an effort, an expenditure of physical and moral energy in order to produce certain results. Even when it is entirely mental, with no words spoken, with scarcely even a gesture; it is still a (voluntary) movement or an attitude of the soul.

Mahalakshmi, a senior professor of biological sciences at IISc, in a similar vein, told me that she does not practise religious rituals like others. She said that she believes in a 'power' but does not believe in the ritualistic aspects of religion. For her, helping others when they are in need is important, and that forms her idea of religion and God:

> "I believe in a power. I am not like many of my family members. They know that I am a scientist, so they don't try to impose their beliefs on me. I do go to temple, but not to worship the idols. I visit temples to enjoy the architecture. Or, I would stand outside the temple and watch people doing various practices. I don't disrespect them when they give me *prasadam*.[6] I do take them. But I don't follow these rituals myself. I don't believe in the dogmatic idea of God. Many people get satisfaction out of various rituals. I don't deny that. But I don't believe in these rituals. The power I believe in is not explainable. For me, if I help someone every day, that is more important than going to the temple. If that power will see my deed, that is enough. Some people get satisfaction and psychological well-being by visiting the temple. But for me, I get satisfaction by helping others. I like to listen to Vedic chants, but I don't chant myself. In that sense, I am not a believer in the traditional sense of the term. For me, temple visit means see the art, architecture and the temple jewels, utensils, etc. For me, God means

> personification of goodness, kindness and helping attitude. As I said, for me doing one good deed daily to anyone around is important. That is my way of practising religion."

A distinguished particle physicist, Geetha Argade, said during the interview that she comes from a traditional Marathi-Saraswat Brahmin background. She stated that she does not believe in the ritualistic part of religion. She said:

> "In my view, ritualistic part or ritual is religion.... You are doing *puja*, going to temple, etc ... that is what I call religion. In that sense, I am not a religious person because I am not ritualistic. For me, religion is a matter of personal choice. Having said that, much of our understanding received from scientific development is not complete. When you study and analyse more on the issue, you will get to know that we haven't understood many phenomena. There is no question of me not believing in that power. As scientists, we think that we understand everything, but we don't. Being a scientist, I know that many things are not understandable. For me, the concept of God has always been associated with something that I don't quite understand. And I believe in that power which is beyond the understanding of nature and science. That doesn't mean that I go to the temple or do *puja* daily. I don't go to temple for my happiness as I don't believe in rituals. However, I visit temple with my parents as it pleases them."

Even though she doesn't believe in rituals and the ritualistic part of religion, she said that she believes in a 'power'. She echoed many other scientists I conversed with. They associate 'dogmatic and ritualistic religion' with the lay believers. As scientists, they think they can be 'spiritual' and not ritualistic. This demarcation between spiritual and ritualistic is one way to differentiate their identity as scientists.

Likewise, Subramanian said, being a particle physicist, it is not difficult for him to see his identity as spiritual as his scientific field also deals with the unknown and the infinite. As he said:

> "I don't go to temple regularly. I don't believe in rituals. Even if I go to temple, I don't go as a believer. I visit temples to enjoy the architecture. It is a pleasure to see that art. I am not ritualistic. I see myself as a spiritual person,

> not as religious. Spirituality means how you relate to the unknown, it is about the infinite. Being a particle physicist, I can associate easily with spirituality. Particle physics also deals with infinite and unknown. So, yes, I see myself as a spiritual person without any doubt."

Being a particle physicist, Subramanian sees himself as a spiritual scientist and says that both particle physics and spirituality deal with questions of infinite and unknown things. I argue that scientists' 'spiritual' identity functions to differentiate them from lay believers and give them a distinct identity as 'scientist-believers'.

Spirituality, like religion, emerged as a category in the West. In his book *The modern spirit of Asia: The spiritual and the secular in China and India*, Peter van der Veer (2014) writes that the term 'spirituality' vaguely alludes to the German term *Geist* ('spirit') and to mysticism. It is a modern Western concept, like 'religion', 'magic' and 'secularity' (Van der Veer, 2014: 35). He argues that the spiritual and the secular were produced simultaneously and in mutual interaction. The modern categories of 'religion', 'secularism' and 'spirituality' were universalized in the age of imperialism and produced different effects in different societies (Van der Veer, 2014: 38). The idea and ideals of spirituality emerged in Europe in response to the onslaught of organized religions as it provided an alternative space. As Van der Veer (2014: 38) argues:

> Perhaps the most important element in the emergence of spirituality was that it offered an alternative to institutionalized religion. Here one sees the connectedness of secularity and spirituality, both limiting religious institutions. In the West spirituality formed an alternative to church Christianity. Together with the so-called secularization of the mind in nineteenth century liberalism, socialism, as well as science (especially Darwin's evolution theory), one can find widespread movements in different parts of the world that search for a universal spirituality that is not bound to any specific tradition.

Dick Houtman has argued that in the context of Western Europe, the traditional language of 'religion' is somehow giving way to a new language of 'spirituality', with many Western Europeans disliking the former and embracing the latter (Houtman, forthcoming: 1; see also Heelas and Woodhead, 2005). He also observes that 'For many Western Europeans, God appears to be no longer a person and a creator who needs to be believed in and obeyed, but rather a non-personal "something" that needs to be experienced in one's own everyday life' (Houtman, forthcoming: 1). Tracing this further, Elaine Ecklund, in her study of American scientists, notes that many scientists see themselves as spiritual but not religious in a traditional sense. She calls this group of scientists 'spiritual entrepreneurs' (Ecklund, 2010: 51). While situated in a different context, this trend has a parallel in the case of these Indian scientists, though, here, spirituality is complexly tied into distinguishing practices.

The narratives discussed earlier clearly demonstrate how scientists try to delineate their idea of God and beliefs differently from lay believers, and the role of concepts of personal spirituality in performing this delineation. Scientists perceived spirituality as a rational system for organizing their beliefs. Their belief in spirituality did not necessarily mean that they stopped practising various traditional rules and rituals that are part of their religion. For instance, the Brahmin scientists continue to practise life-cycle rituals for their children, such as *upanayana* (sacred thread ceremony), naming ceremonies and so on (Thomas, 2016, 2017). Here, as we can see, religious rituals are reduced by the scientists to cultural practices with little transcendental significance.

Another way of being 'distinct' from the lay believer is to show interest in philosophy and literature. For instance, Mahalakshmi, a biologist, told me that she has a good collection of books on temple architecture, and she reads them when she finds time. Susmita Subramanyam (1998), in her study of creativity among IISc scientists, observes a similar trend of scientists being avid readers of philosophy. These were all ways in which scientists differentiated their idea of faith and belief from non-scientists in order to imagine themselves as more rational 'scientist-believers'. They stated that they do not believe in dogmatic religion; rather, they defined and shaped their own ideas of faith and God. Even though they followed the life-cycle rituals and celebrated the religious festivals, they did not wish to be associated with laypeople.

## Questioning compartmentalization

Although many scientists separated religion and science into two different spheres (the first belonging to the private sphere of life and the latter to the public sphere), they did not see them as antithetical. For some of the scientists I met, professional or public life is an extension and continuation of private life. The concept of compartmentalization has often been discussed in several works in Indian sociology and social anthropology. For instance, Milton Singer, in his study of Madras entrepreneurs, argues that the industrialists practised modern (Western) professions without compromising their cultural and traditional way of life:

> This natural history is characterized by the operation of adaptive processes, some of which change the structure of the traditional joint-family system and some of which minimize the conflicts with the new conditions. Among the adaptive processes identified in Madras is the compartmentalization of the domestic and social sphere as 'traditional and religious' and of the industrial sphere as 'modern and secular'. (Singer, 1972: 296)

Similarly, R.S. Khare, in his study of Kanya-Kubja Brahmins, noted that the urban Kanya-Kubja Brahmins try to maintain a balance between their 'home life' and 'office life' without compromising their tradition:

> Thus, as the Kanya-Kubja see it, a modern occupation is not in unmanageable conflict with his home; he does not want to replace one by the other. To him, a modern office [is] at once a symbol of economic necessity, cultural modernity, social status and Brahmanic excellence. (Khare, 1971: 214)

In my research, however, compartmentalization was often questioned.

M.R. Iyer, a renowned scientist and professor emeritus at IISc, decries this compartmentalized way of understanding the science–religion dichotomy. He argued during the interview:

> "I don't believe in compartmentalization and the separation thesis of science and religion. Life is holistic. You cannot say that your science has nothing to do with other aspects of your life. Being spiritual is an important part of doing science. How can anyone compartmentalize it saying that 'This is science, and that is spiritual realm'? Being spiritual helps you to be a better scientist and compartmentalization is not possible in a scientist's life."

Evolutionary biologist and senior professor at IISc Gracy Gomez echoed similar sentiments:

> "There is nothing like personal and professional. It is artificial if someone says that these are separate domains. If you enjoy your work, you cannot treat your profession as a realm confined to your university, and home as your personal realm. These two domains interfere and interact."

Linguist and scholar A.K. Ramanujan (1989) discusses the case of his father, a mathematician and an astrologer, who could easily traverse both his private and public lives without rejecting either. He writes that 'The new ways of thought and behaviours do not replace but live along with older "religious" ways' (Ramanujan, 1989: 57). Similarly, for these scientists, their public and private lives did not oppose each other; rather, they coexisted in a non-antagonistic manner.

## Nostalgia and the revival of science

The scientists' belief in science and religion was also conjoined by a sense of nostalgia about the golden past of Indian science. "All science is in the Upanishads", "Biotechnological research requires ethical principles, which can be derived from the knowledge of ancient India" and "Eastern religions are known for higher levels of thinking" are some opinions of scientists I met on India's golden past of science and technology.

Although they argued that modern Western science is well developed, they viewed ancient India's contribution to science and technology as very significant. Priya Nair, a senior scientist who specializes in cancer research, is proud of the fact that India was at the forefront of sciences in ancient times:

> "Vedic science, concept of zero, *ayurveda*, yoga and meditation are all the contributions of ancient India to modern science. From the Vedic civilization, the foundation of Hinduism was laid. *Ayurveda* is developed by Vedic saints and *Jyotish*. With the help of religion, science was slowly developed in ancient India. *Ayurveda* is a gift given to us by our ancient cultures. We should teach our youngsters to come nearer to nature, understand it and imbibe the simplicity of nature, and *ayurveda* is the best way for this."

Ashok Baruah, who teaches geometry at IISc, told me that the contributions of ancient India to mathematics are very significant:

> "The discovery of positional arithmetic (popularly known as the 'discovery of zero') had a revolutionary effect on our ability to manipulate numbers. I don't think there is any other contribution that comes remotely close to the above. Astronomy of the ancient and medieval era is still relevant in modern science, but most of the breakthroughs were made by the Arabs."

Likewise, Venkataraman, who teaches applied chemistry, emphasized:

> "Contributions of ancient India are enormous to science. Biotechnological research requires ethical principles, which can be derived from the knowledge of ancient India. The successful application of science and technology in society requires careful and methodological approaches that can be developed by studying ancient culture. In ancient India, religion and science were taught together in union, not in differentiation. Therefore, the relationship was very good."

Chandrasekharan is a renowned physicist who serves as emeritus professor at IISc. He comes from a Nambudiri (Brahmin) family in Kerala. He said that he does not follow any traditional rituals and symbols related to his community, including wearing *poonul* (sacred thread). He stated that India had a great tradition of knowledge and learning:

> "We had a great tradition in mathematics. Kerala, for example, had a very strong school of mathematics. We have championed in various fields such as astronomy, *ayurveda* and many others. We only need to look at our old texts such as *Charakasamhita, Vishwasamhita* to understand the rich tradition. But the right-wing argument that we had everything in India is not at all true, that is nonsense.[7] There is no doubt that we had a strong tradition in science. The problem was that we did not have an interpreter. We didn't have a Needham.[8] We have to objectively examine our old great traditions."

These narratives provide a variegated picture of how the scientists dealt with their tradition and science. Some suggested that Indian texts had all the answers to modern scientific problems, while others claimed the superiority of Hinduism. These narratives not only mirror their obvious agreements and disagreements with the scientific authority of the Hindu scriptures, but also reveal the ways in which culture and religion were used interchangeably to invent an exclusive tradition. Historians of science show that India's contributions to science and technology were, indeed, very significant (see Joseph, 2009, 2010), but one needs to clearly delineate this from the ways in which cultural nationalists employ such history to assert the dominance of a Hindu past.[9] One needs to clearly distinguish between the *history of science* from a *nostalgia for science*. As anthropologists and sociologists, we need to be alert to the ideological construction of tradition, which is premised upon a selective recovery of a Hindu past and used largely to assert communitarian claims over the nation-state.

## Scientists' views and wider conflict on campus

During my time at IISc, various organizations and associations organized talks on Hindu contributions to science and technology. I attended talks organized by right-wing student associations and other cultural associations. *Praharshini* and *Prasthutha* are student organizations that organize talks and programmes on Hinduism, Sanskrit, Vedas and Vedic sciences by religious leaders, scientists and various right-wing scholars. I attended one of their talks by the Belgian Indologist Koenraad Elst on Vedic sciences and Indian tradition. The talk was attended by many students and members of the faculty. The student council and the alumni association of IISc also organized talks on the contribution of Hindu cultures in the field of science and technology. One such talk organized by the IISc Alumni Association (IIScAA) was titled 'Epistemological aspects for science: an Indian perspective',[10] by Veda Varidhi Dr P. Ramanujan. Ramanujan is the director of Centre for Development of Advanced Computing (C-DAC), Bangalore. He studied computer science at IISc and also studied Vedas. The talk was moderated by Professor V.V.S. Sarma, honorary professor at the Department of Computer Science and Automation, IISc. It was attended by very renowned IISc scientists, young scientists and researchers. Many such talks and lectures organized by right-wing groups at IISc attempted to show how great and significant the scientific contributions of the Hindu tradition were. They attempted to demonstrate the superiority of Hinduism over other religions. In that process, these talks and lectures at IISc promoted cultural nationalism.

Tata Book House is the only bookshop on the campus. The bookshop has a huge clientele, from senior professors to students of IISc. Here, one finds all advanced scientific books and texts. One also finds books on Vedic mathematics, *Mahabharata*, *Ramayana* and consciousness, alongside texts by Richard Dawkins. When Concern[11] organized a talk by historian of science and critic of Hindutva Meera Nanda, it was attended by more than 100 students. Whenever she criticized Hindutva cultural nationalism, the right-wing students asked questions and counter-questions. The talk resulted in a heated debate regarding the credentials required for a scholar to engage with ancient texts. It also generated extreme reactions from students who not only professed a pride in historical scientific traditions, but also were wary of criticisms challenging its authority.

In September 2012, I conducted participant observation at a two-day conference organized by the Bhaktivedanta Institute (BVI) held in J.N. Tata Auditorium, IISc. BVI is a wing of the International Society for Krishna Consciousness (ISKCON). They organize talks and seminars on science and spirituality in various institutions and an annual conference. They also run small libraries where one can pick up books on religion and spirituality written by world-renowned scientists, as well as mobile libraries in specially decorated buses called 'Science and Religion Library'. They publish books on science and spirituality written by world-renowned scientists[12] and religious leaders. In their publications, they show how various Western scientists talk about Indian scriptures and ancient science. The annual conference is called the All India Students' Conference on Science and Spiritual Quest (AISSQ). The BVI organize the AISSQ conferences in various science and technological institutions.

IISc was host to the seventh session of AISSQ in 2012. It was attended by students of science and technology from different parts of India. Many renowned scientists, including the deputy director of IISc, were present and spoke about the role of spirituality in life. Most scientists spoke on the greatness of Indian tradition and science in ancient India. They spoke on issues such as Vedantic evolution, consciousness, the comparison of modern physics with Hindu science, the importance of vegetarianism and the role of spirituality. When I asked one of the BVI activists about this conference, he told me that their aim is "[T]o attract students in science and technology and empower them spiritually as they are the leaders of tomorrow".

The BVI's founder director, Dr T.D. Singh (1937–2006), envisaged the conference. He was a disciple of Swami Prabhupada, the founder of ISKCON.[13] Dr Singh was also a scientist with a PhD in physical organic chemistry from the University of California, Irvine, US (Mishra et al, 2012). I attended some of the other talks organized by BVI in IISc. They talked about the importance of Hindu tradition, how modern science originated from the ancient Hindu tradition and so on. Most of their talks romanticized Hindu tradition and uncritically praised science in ancient Indian scriptures and texts. In August 2014, I received a mail from friends at IISc about a quiz competition on the Vedas run by the BVI in IISc. IISc generally advertises their academic programmes and events through Broadcast-IISc, an email service for the students and faculty of IISc. However, this time, they used this service to promote a 'Vedic Quiz Competition' as part of the Janmashtami celebration.[14]

One can argue that, in addition to the foregoing examples, the consolidation and culturing of science was also evident in scientists' views on naming satellites after Hindu scriptures or legends from Hindu texts. Termed 'cultural' and not antithetical to the principles of Indian secularism, a leading scientist said:

> "I don't think Hindu naming of satellites is against the ethos of Indian secularism. Most planets have been named after Greek gods. It is natural to do so. Hindu names used for satellites are part of the cultural tradition shared by all Indians, irrespective of their current religious affiliations."

Many scientists echoed this view. Some talked about a shared cultural tradition, which meant that all Indians were essentially Hindus, or share Hindu culture, "irrespective of their current religious affiliations". Therefore, when they talk about various festivals related to Hinduism, they call it 'cultural' (see Thomas and Geraci, 2018). This distinction between culture and religion, I argue, is a way to depict certain practices as 'authentic' Indian practices. Celebrating and perceiving Hindu festivals as 'cultural' is not uncommon in India, and the IISc is no different in this regard. As noted earlier, scientists make a distinction between their belief in God and the belief of the non-scientists in a distinct identity. It is also to be noted that these scientists perceive many Hindu festivals and rituals as 'cultural' and practise them without any discomfort. The normalization of ritual practices as cultural legitimates some traditions at the expense of others. As we have observed elsewhere:

> Identifying one set of practices as cultural lends legitimacy to those practices and also, perhaps, to practices affiliated with them. Many Hindu scientists prefer to believe that festivals such as *Deepavali*, *Holi*, *Ganesha Chaturthi*, and *Ayudha Puja* are cultural festivals. Unless Christian or Muslim,

> however, IISc's scientists do not celebrate or participate in Christmas or Eid (for example), nor does the institute officially recognize these events in any substantive manner. In contrast, the culturalization of festivals like *Ayudha Puja* can normalize the majoritarian religion or tradition. (Thomas and Geraci, 2018: 116)

## Conclusion: studying science and religion in India

This essay has examined how Indian scientists construct a distinct identity as scientist-believers in order to differentiate them from lay believers. It also discussed how they dealt with questions of tradition and culture, and how their answers to these questions feed into wider conflicts over Hinduism in Indian society. While it is important to argue that the Western-centric conflict thesis of science and religion is not universal, and different cultures will receive both science and religion differently, it is also important to note that when notions of Indian tradition and culture are discussed, it is always the tradition and culture of the dominant community that are perceived as authentic Indian culture and tradition. In the case of IISc, as noted, most of the scientists are Brahmins and the culture and tradition that they are talking about is not free from their caste affiliation as Brahmins. When we study religion and science in the Indian context, while acknowledging the particularity of the Western discourse of science and religion, we should also be careful not to reinforce a cultural nationalist idea of science and religion – which would say, for example, that Hinduism and science naturally coexist – because such an idea is historically inaccurate. Sociologists and historians of science should carefully ask: what culture/tradition and whose culture/tradition are we talking about? We should be careful in not generalizing the experiences of the dominant community – in the context of India, Brahmanical Hinduism – to understand science and religion in a particular national context.

### Acknowledgements

I would like to thank Stephen Jones, Rebecca Catto and Tom Kaden for inviting me to contribute to this collection. I am thankful to Susan Visvanathan for detailed comments and valuable suggestions on an earlier version that helped me in the process of writing this chapter.

## Notes

1 The usage of 'Indian culture and tradition' connotes the upper-caste Hindu Brahmanical culture as the majority of the scientists at IISc are Brahmins. For a detailed discussion on science and caste in India, see Sur (2011), Thomas (2015) and Subramanian (2015).

2 It is important to note that reference to God and discussion of religion are not uncommon in the autobiographies of Indian scientists (see Ramanna, 1991; Rao, 2010; also see Thomas, 2018). Here, I am specifically looking at their acknowledgement of God in scientific publications.

3 The Upanishads are Sanskrit texts that are central to Hinduism, and the Brihadaranyaka Upanishad is among the oldest and most significant.

4 Names have been changed to maintain anonymity.

5 *Puja* refers to prayer rituals in Hinduism.

6 *Prasadam* is simple food that has been ritually offered to the gods at a temple.

7 The Hindu Right asserts that Hinduism is somehow uniquely close to scientific investigation. This so-called coexistence of science and religion in Hindu tradition has been criticized and questioned by various scholars (Nanda, 2004, 2010; Brown, 2012). In contemporary times, the right-wing Hindutva groups in India argue that all the sciences originated in Hindu traditions. They also argue that Hinduism was never in conflict with science, and that the theory of evolution was compatible with Hinduism (Nanda, 2010; Brown, 2012). They believe that the ancient Indian traditions had the answers to questions that modern natural sciences have asked or will ask in the future (Nanda, 2010: 280).

8 He was referring to Joseph Needham.

9 For a parallel discussion in the Islamic context, see Bigliardi (2017).

10 Held on 22 September 2012, 4 pm, Faculty Hall, IISc.

11 Concern is a left-wing student group. I helped Concern organize this talk. I was interested in knowing how students in general respond to her criticism of Vedic sciences and Hindutva cultural nationalism. Nanda gave a talk titled as 'The reception of Darwinism in India: what it says about our views of science' on 17 December 2012, 6 pm, Choksi Hall, IISc.

12 One such book is Singh (2004).

13 For a detailed discussion on ISKCON's construction of Vedic science and ISKCON's attitudes towards modern science, see Nanda (2004) and King (2015).

14 A Hindu festival that celebrates the birth of Lord Krishna.

## References

Bharathy, M. (2006) 'Layered oxides and phosphates of bismuth: new structural types and related properties', PhD thesis, IISc, India.

Bellah, R.N. (1971) 'The historical background of unbelief', in R. Caporale and A. Grumelli (eds) *The culture of unbelief*, Berkeley, CA: The University of California Press, pp 39–52.

Bhaskaran, C. (1997) 'Insertion reactions of copper (I) aryl oxides with heterocumulenes', PhD thesis, IISc, India.

Bigliardi, S. (2017) 'The "scientific miracle of the Qur'ān," pseudoscience, and conspiracism', *Zygon*, 52(1): 146–71.

Brown, C.M. (2012) *Hindu perspectives on evolution: Darwin, dharma, and design*, New York, NY: Routledge.

Chacko, R.T. (2007) 'Functionalization of hyperbranched polythers', PhD thesis, IISc, India.

Cimino, R. (2014) *Mystical science and practical religion: Muslim, Hindu and Sikh discourse on science and technology*, Lanham, MD: Lexington Books.

Ecklund, E.H. (2010) *Science vs religion: What scientists really think*, New York, NY: Oxford University Press.

Heelas, P. and Woodhead, L. (2005) *The spiritual revolution: Why religion is giving way to spirituality*, Oxford: Wiley-Blackwell.

Helmreich, S. (1997) 'The spiritual in artificial life: recombining science and religion in a computational culture medium', *Science as Culture*, 6(3): 363–95.

Houtman, D. (forthcoming) 'Pure religion and real sacrality: the cultural logic of an anti-institutional religious tradition', in R. Hefner, C. Timmerman, L. Braeckmans and S. Mels (eds) *Religion and culture in a globalized world: New paradigms and research*, Berlin: Peter Lang.

Joseph, G.G. (2009) *A passage to infinity: Medieval mathematics from Kerala and its impact*, New Delhi: SAGE Publications.

Joseph, G.G. (2010) *The crest of the peacock: Non-European roots of mathematics* (3rd edn), Princeton, NJ: Princeton University Press.

Khare, R.S. (1971) 'Home and office: some trends of modernization among the Kanya-Kubja Brahmans', *Comparative Studies in Society and History*, 13(2): 196–216.

King, A.S. (2015) 'Vedic science, modern science and reason', in I. Keul (ed) *Asian religions, technology and science*, New York, NY: Routledge, pp 54–73.

Mauss, M. (2003) *On prayer*, Oxford: Berghahn Books.

Mishra, S.C., Ghosh, S. and Agarwal, V. (2012) *Science and spiritual quest: Proceedings of the 7th AISSQ conference*, Kolkata: Bhaktivedanta Institute.

Nanda, M. (2004) *Prophets facing backward: Postmodern critiques of science and Hindu nationalism in India*, New Delhi: Permanent Black.

Nanda, M. (2010) 'Madame Blavatsky's children: modern Hindu encounters with Darwinism', in J.R. Lewis and O. Hammer (eds) *Handbook of religion and the authority of science*, Leiden: Brill, pp 279–344.

Nayak, S.K. (2010) 'Dosorder-polymorphism and co-crystal formation in molecular crystals: an in-depth study in terms of weak intra and intermolecular interactions', PhD thesis, IISc, India.

Numbers, R.L. (2007) 'Science, secularization and privatization', in N.A. Rupke (ed) *Eminent lives in twentieth-century science and religion*, Frankfurt: Peter Lang, pp 235–48.

Padmanabhan, T. (2001) *Theoretical astrophysics: Volume II stars and stellar systems*, Cambridge: Cambridge University Press.

Ramanna, R. (1991) *Years of pilgrimage: An autobiography*, New Delhi: Viking.

Ramanujan, A.K. (1989) 'Is there an Indian way of thinking? An informal essay', *Contributions to Indian Sociology*, 23(41): 41–58.

Rao, C.N.R. (2010) *Climbing the limitless ladder: A life in chemistry*, New Jersey, NJ, and Singapore: IISc Press and World Scientific.

Singer, M. (1972) *When a great tradition modernizes: An anthropological approach to Indian civilization*, Delhi: Vikas Publishing House.

Singh, T.D. (2004) *Seven Nobel laureates on science and spirituality*, Kolkata: Bhaktivedanta Institute.

Subramanian, A. (2015) 'Making merit: the Indian institutes of technology and the social life of caste', *Comparative Studies in Society and History*, 57(2): 291–322.

Subramanyam, S. (1998) 'A psychological study of creativity among Indian scientists', unpublished PhD dissertation, National Institute of Advanced Studies (NIAS), Bangalore.

Sur, A. (2011) *Dispersed radiance: Caste, gender, and modern science in India*, New Delhi: Navayana.

Thomas, R. (2015) 'Religious and scientific imagination: a study of religious life of the scientific community in India', unpublished PhD dissertation, Jawaharlal Nehru University, New Delhi.

Thomas, R. (2016) 'Being religious, being scientific: science, religion and atheism in contemporary India', in Y. Fehige (ed) *Science and religion: East and West*, New York, NY: Routledge, pp 140–57.

Thomas, R. (2017) 'Atheism and unbelief among Indian scientists: towards an anthropology of atheism(s)', *Society and Culture in South Asia*, 3(1): 45–67.

Thomas, R. (2018) 'Beyond conflict and complementarity: science and religion in contemporary India', *Science, Technology and Society*, 23(1): 47–64.

Thomas, R. and Geraci, R.M. (2018) 'Religious rites and scientific communities: *Ayudha puja* as "culture" at the Indian Institute of Science', *Zygon: Journal of Religion and Science*, 53(1): 95–122.

Van der Veer, P. (2014) *The modern spirit of Asia: The spiritual and the secular in China and India*, Delhi: Orient Black Swan.

7

# Muslim Perceptions of Biological Evolution: A Critical Review of Quantitative and Qualitative Research

*Jessica Carlisle, Salman Hameed and Fern Elsdon-Baker*

## Introduction

This chapter provides a critical overview of quantitative surveys and qualitative research on Muslim publics' understandings of, and attitudes to, evolutionary science. This topic has received increasing attention over the past decade at the margins of science and technology studies, especially in the overlapping fields of the public understanding of science (PUS), science communication and science and society. Research into publics' understanding of science typically crosses over a number of research fields and includes data collected by special interest groups (eg science policymakers and educators). This is especially the case for research into Muslims' understanding of science, which is often found in studies interested primarily in the role of religion in society. Our review thus incorporates small-sample, interview-based qualitative studies and larger quantitative surveys across many related fields. In what follows, however, we pay particular attention to how Muslim acceptance of evolution features in PUS.

As a field of study, PUS has well-established lexicons, research questions and hypotheses for investigating publics' attitudes to evolution. In Muslim contexts, despite the limited number of empirical studies, there is an entrenched assumption in PUS research that the rejection of evolutionary science by Muslims is a widespread problem resulting from these publics' faith in Islam. This chapter considers the assumptions underpinning most research, discusses the limits that this imposes on findings and highlights the insights brought by some studies adopting an open, exploratory approach to Muslim experiences and opinions. Most existing studies have serious shortcomings: foregrounding an essentialized Islam; ignoring plurality and diversity across different societies; and ignoring the lived experiences of being Muslim. In conclusion, we suggest ways of dealing with these issues and improving research on Muslim publics' understandings of evolutionary science.

## The importance of knowledge deficits in PUS research on biological evolution

Since the 1980s, as public attitudes towards biological evolution became subject to growing scrutiny, PUS and latterly science and society[1] research has been allied with the mainstream scientific community in assuming that public acceptance of evolutionary science is important, or, indeed, a form of social good. This project is rooted in a foundational assumption in facets of PUS literature that the relationship between science and society is fractured (Bodmer, 1985; Suldovsy, 2016). There is, in Bauer et al's (2007: 90) words, 'a common sense prejudice among experts' that a deficit in the 'public understanding of science' (including ignorance about evolutionary science) exists and is detrimental to societies. The solution offered by PUS and science educators follows this 'deficit' model of public understanding and emphasizes the better communication of scientific methods and findings, including those of biological evolution.

The prominence of evolution in PUS is arguably, in part, due to American influence. The doctrine of creationism acquired profound cultural status in the US during the 20th century, informed by political power struggles between federal and state government, a series of moral crises stoking a culture war, and public debate about the role of faith (predominantly Christianity) in government and in public schools. The theory of biological evolution – especially the origin of humans – became a bellwether in the second half of the 20th century for US public opinion on scientific methods and findings, informed by a politics in which scepticism about science, articulations of Protestant faith (including concerns about scriptural literalism) and conservative political ideology combined and found expression in political campaigns and high-profile court cases involving the teaching of creationism in public schools (Webb, 1994; Larson, 2007; Wiles, 2011; Shapiro, 2015).

Studies of public understanding of evolution in the US have, consequently, concentrated on identifying which publics are particularly resistant to evolutionary science, explaining how publics justify their rejection of it and identifying what factors would improve public understanding. These issues have been addressed by philosophers (Pennock, 2000; Ruse, 2005), historians (Numbers, 2006; Moran, 2012), scientists and educators (Berkman and Plutzer, 2010; Coyne, 2012; Rosengren et al, 2012), and those who study social movements (Eve and Harrold, 1991; Lienesch, 2007).

A central finding of this research in the US is that an individual's or a community's religious faith is strongly correlated with a rejection of evolutionary science (see Hill, Chapter 2, this volume). Some of these studies have produced taxonomies of how religious identity predicts attitudes towards evolutionary science (with US Catholics being much more likely to accept evolutionary science than evangelical or fundamentalist Christians). This research finding underpins a fundamental supposition in studies of global publics' understandings of evolution that individuals and communities with faith are more likely to reject evolutionary science than non-religious individuals or communities. This analysis further assumes that individuals of faith remain influenced by socialization that has made them dismissive of (particularly human) evolution, and that communities replicate their belief in creationism through refusing to engage with scientific argument (including challenging the supremacy of evolutionary science over creationism in schools).

This academic discourse has been reinforced by political, legal and social debates in the US. There have been a number of high-profile court cases – notably, involving the US Supreme Court in 1987 (see Kaden et al, Chapter 3, this volume) – which have addressed the status of 'creation science' or its later incarnation, intelligent design, as a valid scientific idea to be taught in public schools (Berkman and Plutzer, 2010). There is an ongoing and continuous struggle at the level of local school boards in the US on the presentation of evolution in biology textbooks, including efforts – occasionally successful – by some to add disclaimers about evolutionary science.[2] While there are complex social, religious and political factors behind the creationist movement (Eve and Harrold, 1991; Lienesch, 2007), many scientists and science educators in the US have come to view resistance to biological evolution as a fundamental challenge to the status of science.

Consequently, a person's acceptance of evolution has become equated with their acceptance of science.

An indication of the importance of this issue for elite US professional scientists is indicated by the inclusion, since 1983, of public attitudes to evolution as a measure of the public's understanding of science in the US's influential Science and Engineering Indicators (SEIs). In 2010, arguing that the evolution question (along with a question about the Big Bang) may measure more about people's beliefs rather than their knowledge, the question was dropped as a measure of what people 'know' about science. This drew criticism from some scientists, who accused the SEIs for hiding a 'national embarrassment', with some even calling the effort an 'intellectual malpractice' (Bhattacharjee, 2010).

Although the topic of biological evolution has been relatively less politically charged outside of the US throughout the late 20th century, international scientific associations (Stears et al, 2016), the Council of Europe (Hameed, 2015) and national science academies and networks (BouJaoude et al, 2011a) have all responded to concerns about a perceived global rise of creationism. Outside the US, creationism has consequently attracted increasing media attention, including examples in Muslim-majority contexts. By the end of the 2000s, for instance, the Turkish creationist Harun Yayha (Adnan Oktar) was receiving coverage from the US and European media. In 2011, the Education Committee of the Council of Europe cited Yahya's Islamic creationism, among other creationists, as one of the threats to global education. Subsequently, some academics have argued that creationist ideas and frames were being exported out of the US to regions including the Middle East, where they have 'a certain, and dangerous, resonance with an apparently large portion of the Muslim world' (Wiles, 2011; for a history of creationism in the US and its global impact, see also Numbers, 2006). Similarly, a statement by the InterAcademy Panel on International Issues (IAP, 2006), to which a number of national academies (including those of many Muslim-majority states) were signatories, singled out the issue of human evolution as crucial for science education.

PUS has broadly become more reflexive in the last couple of decades in its efforts to understand public perceptions of science in the West (Bauer et al, 2007; Kahan, 2015), and this reflexive approach has started to include the study of public perceptions of biological evolution (Evans and Evans, 2010; Hill, 2014; Elsdon-Baker, 2018). However, most studies of Muslim-majority and -minority publics, as we will see across this chapter, continue to maintain a deficit model approach, replicating the critique that religion functions as a particular obstacle to the public understanding and acceptance of evolution (Numbers, 2006; Coyne, 2012; Moran, 2012; Paz-y-Miño and Espinosa, 2015).

## Quantitative studies of Muslim attitudes to evolution: proving a point

Until 2006, only a few small studies had analysed perceptions of evolution in a Muslim context (eg Dagher and BouJaoude, 1997). However, increased interest in the topic provoked a number of large, global comparative surveys including Muslim respondents. In what has become one of the most-cited comparative surveys of global perceptions of biological evolution, Miller et al (2006) analysed survey data on responses in the US, Japan and 32 European countries to the assertion: 'Human beings, as we know them, developed from earlier species of animals'. The results were interpreted by the research team as confirming that conservative or fundamentalist religious faith was a barrier to full acceptance of biological evolution.

Over 80% of respondents in Iceland, Denmark and Sweden agreed with the statement, in contrast to 40% of respondents in the US and 25% in Turkey. In their publicity surrounding their analysis of these statistics, the authors noted that 'only' the general public in Turkey, which is described as 'a predominantly Muslim nation' (Michigan State University, 2006), were less supportive of evolutionary science than the US population. In media coverage reporting these results, Miller subsequently extrapolated wide-reaching generalizations about Turkish society and religious identity by claiming that US Protestantism and Turkish Islam were both 'fundamentalist' faiths (quoted in Live Science, 2006). While this study is one of the most widely cited in this field, it is important to note that the structural equation model on which the majority of the more detailed findings are based (including those for the US) does not include data for Turkey. These comments were thus only based on what the authors term 'descriptive analysis' of a single-item measure for the 2005 Eurobarometer survey undertaken in the 32 countries.

In ascribing the Turkish public's resistance to evolutionary science to fundamentalism and associating it with evangelical Christianity, Miller squeezes this descriptive analysis of a single survey question into a framework informed by PUS analyses of the US and the 'deficit model'. In the supporting materials, the authors recognize that Turkey is 'a country still debating the issue of secular versus theocratic government' (Michigan State University, 2006). However, their comparative analysis, as with most other overviews of this kind, did not sufficiently place their findings in the Turkish political and social context and drew on stereotypes in its description of Turkish religiosity. In 2006, the Justice and Development Party (AKP)-dominated government of the moderate Islamist and democratic nationalist Recep Tayyip Erdogan had been in power for four years, and Erdogan was only just beginning his campaign to confront the Turkish military command and higher judiciary committed to the constitutional principle of secularism (Article 2 of the 1982 Constitution). Turkey's recent economic boom has lifted a large contingent of the population into the middle class and created a consumer society. Religious identity among Turkish Muslims is accordingly complex, with a high percentage of respondents replying to some surveys by stating that they are atheist or agnostic. Moreover, Sufism and the Gülen Movement (Tee, 2016) are an important aspect of many Turkish Sunni Muslims' identity. This complexity is not reflected in the characterization of the Turkish general public as 'fundamentalist'.

Hassan's (2007) subsequent contribution to representations of Muslim rejection of evolution analysed religious patterns among the 'highly educated groups occupying professional, economic, social, religious, cultural and bureaucratic positions in the mainstream social structures' in Egypt, Indonesia, Kazakhstan, Malaysia, Pakistan and Turkey.[3] This study of six Muslim-majority countries included a question about evolution on the assumption that evolutionary science challenges a 'fundamental religious belief widely held by Muslims' (Hassan, 2007: 465).

The question read: 'Do you agree or disagree with Darwin's theory of evolution?'. This survey item is problematic for a number of reasons. First, it assumes a working knowledge of Darwinian evolution. Second, it uses the word *theory*, which could be perceived by members of the public as implying uncertainty, regardless of how well established evolution science may be within the scientific community. Third, it historicizes a contemporary field of research by linking it with Darwin. Thus, this study is indifferent to the potential negative associations between the name 'Darwin' and an assertion of Western, scientific supremacy/materialism (as explored in more detail later in the chapter). Finally, this question did not differentiate between human and non-human evolution. A recent study in the UK and Canada has shown that aspects of human evolution (as opposed to evolution of other life forms, including animals) is the most common issue or concern raised by survey respondents across all groups (Elsdon-Baker et al, 2017). It is therefore important when undertaking this kind of survey, as Hill (2014) has also highlighted in the US context, to ascertain exactly what facets of biological evolution are being rejected.

Thus, Hassan's packaging together of human and non-human evolution, along with his use of ambiguous or loaded terms, means that he provides limited insight into what evolution as a term might mean to these publics, or, indeed, exactly what it is that is being rejected. Without more information about the respondent's understanding of the question, it is impossible to interpret too much from these data. While the survey's author acknowledged its limitations, Muslim responses to this question reinforced a popular impression of widespread Muslim opposition to biological evolution (Hameed, 2008).

Additional surveys including respondents in Muslim countries have further highlighted how the problematic phrasing of questions concerning biological evolution makes the results difficult to analyse. For example, in 2009, the British Council 'Darwin Now' project conducted a survey of 10 countries, including Muslim-majority Egypt.[4] The survey found that close to 60% of Egyptian respondents agreed with the statement[5] that: 'Life on earth, including human life, evolved over time in a process guided by a God'. Only 30% expressed the opinion that life on earth was created by God and has 'always existed in its current form' (Elsdon-Baker, 2015). The phrasing of this question, which is not dissimilar to that used by Gallup in their US-based surveys (see Hill, Chapter 2, this volume), is also ambiguous – especially when taken out of context from the US. This style of question is constructed around a US-influenced concept of not only what creationism is and means to publics, but also what we might think of as theistic evolution. Its success also rests on respondents having a good understanding of the meaning and implication of the terms 'origin of species' in the question phrasing and 'evolved' in the item phrasing. However, it is very important to note that when Egyptian respondents were asked, in the same survey, 'Which, if any, of the following statements best describes how well you understand the term "evolution"?', 62% of the respondents said that they had never heard of the term. A further 9% said that they had heard of the term but had no understanding of what it meant. Furthermore, in a follow-on question, 62% also stated that 'I have never heard of Charles Darwin or his theory of evolution', with a further 15% stating that while they had heard of Darwin and his theory, they knew nothing or little about them. Egypt came out lowest of all the 10 countries polled in terms of awareness and understanding of the term 'evolution' (Elsdon-Baker, 2015). We should therefore be very cautious about taking the findings in the 'Darwin Now' poll at face value within the Egyptian context.

Difficulties in phrasing also plague a more recent large-scale survey 'The world's Muslims: religion, politics and society' (Pew Research Center, 2013), which surveyed Muslim respondents in 22 countries. Most of the questions in this survey address religious and political issues. However, the one question that explicitly addresses human evolution asks, 'Thinking about evolution, which comes closer to your view?' and gives respondents a choice between the statements: 'Humans and other living things: a) have evolved over time or b) have existed in their present form since the beginning of time'. Only a stark binary choice is presented, and no further questions make clear if respondents have an awareness or understanding of the term 'evolution'. This question is again very much based on a US-centric model of Young Earth creationism, despite us having little idea of whether this is salient to publics outside of US Protestant Christian contexts. Moreover, the human evolution question comes directly after a question that asks: 'In your opinion, do you think there is generally a conflict between science and religion, or do you not think so?'. It is not inconceivable to think that this juxtaposition might have acted as an influence upon how respondents reacted to the evolution question.

Problems are also embedded within the wording of the question. Notably, it leaves open the possibility that some of the respondents may be envisaging adaptation *within species* when agreeing with the statement 'Humans and other living things have evolved over time'. This is what is commonly referred to in creationist literature as 'micro-evolution'.[6] This model allows for a form of development within the human species but rejects what is sometimes called 'macro-evolution', the idea that humans have developed from another pre-existing species or that we share a common ape-like ancestor with other hominids, including great apes (such as chimpanzees or gorillas). Clearly, a human exceptionalism model that allows only for micro-evolution and not macro-evolution within the human species is not an endorsement of evolutionary science. We have little idea of how these models may play out with Muslim publics, but it is vitally important to design survey questions and items that allow us to build a better understanding of this kind of nuance.

The Pew survey unsurprisingly produced large country-based variations in responses, with Kazakhstan returning the highest level of evolution endorsement (with 79% agreeing with option [a]) and Iraqis responding least positively (with 67% with the 'creationist' option [b]). It was only in Iraq (67%), Afghanistan (62%) and Tajikistan

(55%) where the majority of respondents selected the evolution rejection option. Overall, 53% of Muslim respondents agreed with the statement that humans and other living things have evolved over time. Interestingly, as the Pew report on this survey notes:

> American Muslims are somewhat less likely to believe in evolution than are Muslims in other parts of the world (45% vs. global median of 53%). Indeed, when it comes to evolution, US Muslims are closer to US Christians (46% of whom say they believe in evolution) than they are to fellow Muslims elsewhere in the world. (Pew Research Center, 2013: 36)

This potentially further indicates the amplified cultural significance of these debates in the US in contrast to other cultural contexts.

In addition to these surveys, a number of quantitative studies of science education in Muslim-majority countries have used the Measure of Acceptance and Rejection of the Theory of Evolution (MATE) to assess student attitudes towards evolutionary science with little or no modification to take into account local contexts (Peker et al, 2010; Deniz et al, 2011; Yousuf et al, 2011). Some of these studies are useful for gaining an insight into what students understand about biological evolution and common scientific misconceptions about evolutionary science, but the foundational principle remains a 'deficit' model. The more sophisticated quantitative analyses of the causes of evolution rejection that one can now find in the US Christian context (see Hill, Chapter 2, this volume) remains almost totally absent in Muslim contexts (with very recent notable exceptions: see Unsworth and Voas, 2018; see also Unsworth, Chapter 12, this volume). Few studies ask people if they see correct evolution belief as important, or look into the social causes of rejection. Rather, the rationale for quantitative research to date seems to be that Muslims' support for evolutionary science can be read as a proxy for civilizational progress.

## Qualitative research: comparative studies of Muslim group responses

Most qualitative assessments of Muslims' attitudes towards evolutionary science emulate quantitative research in foregrounding religious identity in their analyses. This is an emerging field and most of the work to date has been published by a handful of scholars. This research has been conducted in nation-states including Turkey, Pakistan,

Tunisia, Canada, Lebanon, Egypt, Senegal, Malaysia, France and the US. It has broadly been interested in how Islam, or being Muslim, determines governments' or publics' resistance to evolutionary science. Researchers have employed textual analysis in macro-level analyses of the treatment of biological evolution in Muslim-majority states' education policies and schools' curricula, while micro-level studies of public attitudes towards evolutionary science in Muslim-majority and -minority contexts have employed surveys, interviews and ethnography. Qualitative analyses extend beyond reporting survey data. Nonetheless, in contrast to US-focused studies of Christians (Noy and O'Brien, 2018), few of these analyses adopt any form of intersectional analysis of individual respondents' attitudes (including their gender, class, situated ethnicity, etc), and, in general, this research only cursorily recognizes 'differing historical experiences and socio-political dynamics' (Burton, 2011) or discusses religious debates about Islam, creationism and evolution.

Additionally, qualitative studies rarely challenge the assumption running through quantitative research that Muslims' doubts about biological evolution are primarily due to a deficit in their understanding of the science underpinned by religious world views that act as a barrier to engagement and thus social progress. Typically, such studies are marked by a focus on formal sites of science education as a countervailing force or a remedial measure. BouJaoude et al (2011a: 980) are fairly typical in noting that evolutionary science is 'globally and overwhelmingly accepted by the scientific community' and should, as such, 'be a central, and noncontroversial, element of biology teaching'. Chan Kit Yok et al (2015), when reporting the aims of the Biohead Citizen project of 2004–08, associate acceptance of evolutionary science with good citizenship.

On this basis, much qualitative research on Muslim attitudes to evolution broadly splits into two types: either presenting a comparative, inter-religious analysis of Muslim and non-Muslim communities on a sliding scale of different denominations' acceptance or rejection of evolution (BouJaoude et al, 2011a; Chan Kit Yok et al, 2015; Clément, 2015; Stears et al, 2016); or presenting an intra-religious typology of how individual Muslims relate to evolutionary science (Aroua et al, 2009; Mansour, 2010; Everhart and Hameed, 2013; Croché, 2015). The bulk of these studies have focused on biology teachers (Asghar, 2013; BouJaoude et al, 2011a; Mansour, 2010; Taşkın, 2014; Chan Kit Yok et al, 2015; Clément, 2015; Croché, 2015; Stears et al, 2016) or students (Hrairi and Coquidé, 2002; Aroua et al, 2009; BouJaoude et al 2011b) since both are assumed to be well informed about evolutionary science and teachers are identified as crucial to general Muslim publics' understandings of evolutionary science.

The conclusions of comparative studies between religious communities typically stratify denominations of Muslims, Christians and other major world faiths from the most to the least likely to accept evolution. Most of these studies find that Sunni Muslims are most resistant to evolution, though each suggests considerable complexity in their Muslim respondents' attitudes to evolutionary science. Chan Kit Yok et al (2015) have found that Muslim teachers in Malaysia are significantly less supportive of evolution than their Christian and Buddhist colleagues. This result has been replicated in Lebanon, where Clément (2015) concluded that Sunni followed by Shia Muslim teachers are more creationist than Orthodox Christian, Druze and finally Catholic teachers. BouJaoude et al (2011a) have equally concluded that Muslim biology teachers and professors are more likely to reject, partially accept or refuse to engage in discussions about evolutionary science than their Druze or Christian colleagues. In summarizing their findings, they observe that 'participants' religious orientations seemed to influence their acceptance of evolution' (BouJaoude et al, 2011a).

The finding that Sunni Muslims are particularly 'creationist', however, has been questioned by Clément (2015: 413), who has argued, after observing that Protestants in Burkina Faso are more supportive of creationist explanations for the origin of life than Muslims, that there is no 'international standard of creationist beliefs linked to the Muslim religion'. Moreover, research concentrating on diversity within Muslim publics finds that individuals either: reject evolution as in necessary conflict with Islam; believe that evolutionary science can be wholly or partly reconciled with their faith; or separate the realms of 'science' from 'religion' (Mansour, 2010; Taşkın, 2014). Asghar et al (2007) note that general publics' responses reflect Muslim intellectuals' refutation, acceptance or adaptation of the idea of evolution.[7] These categorizations reflect Barbour's (1990) typology of religious individuals' conceptions of the relationship between science and religion as conflict, independence (in which religion and science operate in different domains and provide different ways of knowing), dialogue (in which religion and science mutually inform one another) and integration (in which scientific knowledge is regarded as encompassed by religion) (see also Reid, Chapter 4, this volume). Muslim interviewees have identified the science of biological evolution – particularly as it describes human evolution – as especially difficult to reconcile with their faith (Mansour, 2010). This difficulty has been attributed to prohibitions on the teaching of evolutionary science (Aroua et al, 2009), hostility on the part of teachers to its inclusion in curricula (Wiles, 2011) or long-established gaps in science curricula that allow for human exceptionalism (Asghar et al, 2014; Chan Kit Yok et al, 2015).

Pursuing a concern with the implications of a deficit in the understanding of evolutionary science, and the role that educational settings may have in ameliorating this knowledge gap, researchers have been particularly focused on interrogating Muslim biology teachers' and students' beliefs and attitudes. These studies rely on surveys, or a combination of a survey and interviews, with a selection of respondents. Researchers have based their conclusions on diverse data sets: Clément (2015) has analysed responses from 2,130 Muslim teachers in several Muslim-majority states (using questions found in the Biohead Citizen survey); Chan Kit Yok et al (2015) surveyed 204 Malaysian teachers; BouJouade et al (2011b) surveyed 162 Egyptian and 629 Lebanese students, and interviewed 30 Lebanese students; Taşkın (2014) interviewed four students in the US; BouJouade et al (2011a) interviewed 20 Lebanese teachers and seven university professors; Asghar (2013) interviewed 17 Canadian and Pakistani teachers, and conducted a focus group with six additional Pakistani teachers; and Everhart and Hameed (2013) interviewed 23 American doctors of Pakistani descent.

Most of the interview-based studies find that individuals who are resistant to the expectation that they teach or learn evolutionary science modify their behaviour or presentation of self in educational environments by self-censoring public expressions of their opinions (Taşkın, 2014), being equivocal about the status of the science in comparison to religious explanations for the origins of species (Asghar, 2013; Croché, 2015) or instrumentally demonstrating knowledge of biological evolution without conviction (Hrairi and Coquidé, 2002). Muslim respondents have also been found to qualify their acceptance of evolutionary science by endorsing a model that allows for biological evolution as a process occurring under direction from God (Chan Kit Yok et al, 2015) or denying that humans are the result of evolutionary processes that have produced non-human species (BouJouade et al, 2011a). These kinds of strategic responses from teachers have been identified as a particular matter of concern given their influence on students' understanding of evolution (Stears et al, 2016). A few researchers have recommended that curricula enable the pedagogical acknowledgement of religious perspectives (Asghar, 2013; Croché, 2015), while others have strongly argued against the accommodation of faith (Aroua et al, 2009; Taşkın, 2014). Some researchers have also concluded that improvements in the teaching of scientific methods and methodology would facilitate its acceptance by students (Hrairi and Coquidé, 2002; BouJouade et al, 2011a).

## Decentring Islam in explaining Muslim attitudes to evolution

The broad conclusion from this qualitative research is that being Muslim is inherently an obstacle to public understanding of evolutionary science and that this is a problem. These studies have all adopted a foundational assumption in aspects of science communication that science should be well communicated to publics and that non-scientists should recognize its epistemic supremacy (Sudolvsky, 2016). Moreover, these studies tend to emphasize the importance of scientific literacy, despite this approach being largely abandoned by recent scholarship and practice in PUS and science communication in favour of a model of promoting better engagement or reciprocal dialogue between scientists and publics (Bauer et al, 2007; Sudolvsky, 2016). Consequently, much qualitative research frames its analysis of Muslims' attitudes to evolutionary science in quite narrow terms: subjects are required to state and rationalize their opinions. Their responses are frequently extrapolated to a global assessment of (particularly Sunni) Muslim resistance to evolutionary science.

To be clear, we are not arguing against science education per se. Rather, our objection is that this approach pays little or no attention to specific transnational and national historical contexts, does not acknowledge the impact of its methodology on subjects' responses, negates respondents' intersectional (gender, class, ethnic, urban/rural, generational) realities, precludes interviewees from shaping the research agenda, and significantly overestimates publics' (both Muslim and non-Muslim) rational relationship to science. Only a handful of qualitative studies demonstrate the advantages of taking these issues into account.

Although some researchers have explicitly acknowledged the variations between Muslim contexts, in general, most do not interrogate specific histories of the politics of evolutionary science beyond describing variations between states' education curricula (BouJaoude, 2011b). Aroua et al (2009), to give an extreme example, makes a misleading claim that homogenizes the diversity in states' policies regarding the teaching of evolution. In making an argument for the uniqueness of its politics, she asserts that Tunisia is an exception among Muslim countries in not forbidding the teaching of biological evolution (Aroua et al, 2009). In fact, a minority of Muslim-majority states currently prevent the teaching of evolution, principally in the Gulf: Saudi Arabia, Oman, Yemen and Iraq. Muslim-majority countries, including Iran (Burton, 2011), Pakistan (Asghar et al, 2014), Lebanon, Egypt (BouJouade et al, 2011b), Turkey, Syria and Malaysia (Asghar et al, 2014), have included evolutionary science as part of their curricula.[8]

BouJouade et al (2011b) explain the differences in Egyptian and Lebanese education policies in their study of students' perceptions of evolutionary science. They record a considerable difference in Egypt's inclusion of biological evolution in its national secondary school curricula and Lebanon's non-inclusion of evolutionary science in its decentralized education system, although they note that many students may have encountered evolutionary science in textbooks used in schools following the French and American educational systems (though they fail to add that these schools cater for the middle classes and elites). Several researchers note that religious arguments are rarely included in evolution curricula (BouJouade et al, 2011b), but that the evolution of humans is often not addressed, including in Egypt, Malaysia, Syria, Pakistan and Turkey (Asghar et al, 2014).

This tendency to focus on the content of national curricula has facilitated the exclusion of analyses of national and transnational political, social and theological debates about evolutionary science. The impact of these debates may be indicated by the conclusion of BouJoaude et al (2011b) that, despite the significant differences in the treatment of biological evolution in the curricula they were taught, both Egyptian and Lebanese Sunni Muslim students tend to express creationist ideas. Clément (2015: 413, 418), concluding a cross-country survey, has argued that there is 'an absence of interaction between knowledge and values' in Muslim evaluations of evolution, adding that 'economic, political, geographical and historical parameters' have a greater influence on people's attitudes than 'the specific influence of the Muslim religion'. Falade and Bauer (2018) similarly caution against extrapolating too much about the rest of Nigeria from their findings in Lagos. Everhart and Hameed (2013) also find that 'there is a dynamic and fluid interaction between [Pakistani-origin physicians'] scientific views and their own religious and cultural contexts' when thinking about evolution.

Despite these researchers' recognition of the influence of history, economics and politics on Muslims' attitudes to evolutionary science, there is little analysis of to what extent these influences have rendered evolution politically salient for their research subjects. Asghar et al

(2007) have uniquely outlined some of the problems that they encountered in conducting survey research in Indonesia and Pakistan as difficulties in establishing rapport and trust, suspicions that the research agenda was anti-Muslim, and sensitivities arising from survey statements assessing respondents' perceptions of scientific accuracy in the Qur'an. However, the research group treats the politicization of religion (and biological evolution) as an obstacle to be overcome rather than a field of enquiry.

The gains of a focus on the impact of transnational and national politics are shown in Guhin's (2016) rich comparison between conservative Protestant and Muslim students' discussions of evolution in the US. Based on 18 months of participant observation in four schools, Guhin stresses that while evolution was often discussed in Protestant students' Bible classes, he was the one who brought it up in conversations with Muslim students (Guhin, 2016). He argues that the issue of evolution has little *moral salience* for Muslim students in the US, who are more conscious of being subject to negative discourses about Islam and gender, and, consequently, they are less actively engaged in the rejection of evolutionary science than conservative Protestant students.

Guhin's (2016: 168) ethnographic study emphasizes that 'lay acceptance of scientific authority is … pragmatic and situational'. This approach of investigating *how* Muslims think about evolutionary science rejects the deficit model-informed approach of much qualitative studies that attempt to establish what Muslims *really* think about evolutionary science and how to address any subsequent knowledge gap. It is well established that Muslim students frequently seem poorly informed about evolutionary science, including a common misconception that Darwin stated that humans are direct descendants of apes or monkeys (Asghar et al, 2007; BouJaoude, 2011a; Asghar, 2013). Researchers rarely contextualize this by noting that there is a global lack of knowledge about evolutionary science, with analyses generally relating this deficit of knowledge to Islam or being Muslim.

Guillo (2017) is unusual in demonstrating that although non-Muslim French biology students fully agree with evolutionary science, they are as poorly informed about its explanation for the origins and development of life as Muslim Moroccan students, who tend to dismiss or only partially accept evolutionary science. Having demonstrated that French students' support for evolution is principally ideological (signalling allegiance to a modern, scientific world view), Guillo (2017) observes that most (Muslim and non-Muslim) laypeople have 'fuzzy' conceptions of science about which they are generally indifferent, ceding mastery over it to scientists. He argues that in their determination to discover what (Muslim) subjects really think, researchers are resistant to accepting these fuzzy ideas at face value and try to track down 'some latent knowledge, representation, schema, cognitive biases, automatic intuitions [or] folk biology' (Guillo, 2017: 189). This leads to researchers insisting on unequivocal responses to sensitive questions, with the resultant production of artificial schemas intended to explain 'Muslim thinking' or socialization.

These schemas generally pay little attention to the wide diversity found across and within Muslim publics beyond some consideration of gender (Taşkın, 2014) and, tangentially, socio-economic class (Croché, 2015). This lack of an intersectional analysis of Muslim interviewees' subjectivities is exhibited in Clément's (2015) passing observation that although female biology teachers are more knowledgeable about evolution than their male colleagues, they are, nevertheless, more sceptical about evolutionary science. In failing to analyse individual Muslims' positions and perspectives, researchers have also not interrogated *why* the acceptance of evolution by Muslim, or, indeed, any general, publics is important for different individuals or society (beyond the specific circumstances of biology teachers and students), and what real-world value accepting or rejecting evolutionary science really has in either Muslim, or non-Muslim, contexts.

This lack of analysis of the implications of gender, ethnicity and class sharply contrasts with recent ethnographies of religious and medical perspectives on sensitive topics such as organ donation in Egypt (Hamdy, 2013) and assisted reproductive technologies among Muslims in Lebanon (Clark, 2009) and in Cairo (Inhorn, 2004). These studies address issues that are more ethically problematic for their subjects than for their English-speaking authors/readership, but that have not acquired the totemic and contested status of evolutionary science as a marker of 'global' publics' understanding of science. Moreover, as anthropologists, their authors situate themselves firmly within the social, religious, legal and political discourses that produce individual Muslim's attitudes and decisions regarding in vitro fertilization (IVF)/ intracytoplasmic sperm injection (ICSI) or organ transplantation. The outputs are detailed, non-essentializing descriptions of the fluid intellectual, social and emotional processes contributing to the formation of Muslims' private and public opinions and actions. In particular, these studies show how individual attitudes, expressed in religious terms, fluctuate over time, particularly when citizens in the Middle East or North Africa are faced with a personal crisis such as being in a childless marriage (Inhorn, 2004; Clark, 2009) or having a critically ill spouse (Hamdy, 2013).

## Conclusion: recommendations for future research

The number of academic studies on Muslim publics' perceptions of, or attitudes towards, evolutionary science is striking in comparison to Muslim publics' understandings of, and attitudes towards, other non-medical fields of science. This very narrow focus betrays the origins of a research agenda rooted in the politicization of evolutionary science in the US. Moreover, most studies of this topic miss the opportunity to explore the complexity and diversity of Muslim viewpoints. Most of this quantitative and qualitative research either ranks acceptance of (human and non-human) evolution in a hierarchy of sectarian affiliations, or offers a typology of 'Muslim responses' to it (ie rejection, accommodation or acceptance). The importance of accepting evolutionary science is usually implicit in these studies, while the significance of religion as an explanatory frame is explicit. This is problematic given the lack of analysis of what Muslim research subjects understand about evolution science and what value it has to their lives, world views or societies.

Instead, Islam, or being Muslim, is often predicated as an obstacle that threatens to hinder the understanding and acceptance of evolution. This analysis lacks the reflexivity that is beginning to be found in the wider fields of PUS, science communication and science and society. There is little analysis of the specific socio-political environment of the research field, intersectional recognition of Muslims' heterogeneity (Hameed, 2015) beyond observing that there are different trends between Muslim-majority states (Stears et al, 2016), or recognition of the messy, fluid, situated ways in which Muslims and non-Muslims think and feel about science.

This lack of intersectionality, socio-political context or interest in the mind results in generalized representations of Muslims that fix subjects' attitudes in the moment of enquiry. Islam and being Muslim are frequently evaluated from a perspective grounded in the deficit model's commitment to the supremacy of epistemic concerns. In contrast, ethnographic studies demonstrate that Muslims' thinking about evolution (Guhin, 2016) is as constructed and adaptive as it is in response to such 'real-life' religio-ethical dilemmas as IVF (Inhorn, 2004; Clark, 2009) and organ transplantation (Hamdy, 2013).

This turn towards participant observation in studies of Muslim-majority and Muslim-minority attitudes towards science on issues including procreation, life-saving treatment and human evolution problematizes the notion of epistemic conflict between science and religion by charting the multiple factors informing Muslims' attitudes.

This work suggests that fundamental suppositions about faith in the study of Muslims' attitudes to, and understandings of, evolution should be a subject of reflection.

Consequently, future quantitative research should: be attentive to how it words survey questions; extend its evaluations to Muslim respondents' interest in, and understanding of, evolutionary science; and focus on the political, social, cultural and moral salience of evolutionary science in comparison to other medical and scientific issues. It should also be cautious about making global inferences about 'Islam' and 'Muslims'. Future qualitative research should examine the (political, gendered, class, etc) power struggles and debates within Muslim publics as they think about evolution during situated engagements in politics and social debate, rather than assuming that resistance to evolution is 'merely articulating an expression of Islam' (Tayob, 2012: 161).

## Acknowledgement

The authors would like to thank Sarah Weirich for her contribution to earlier versions of this chapter.

## Notes

1 Bauer et al (2007) identify 'science and society' as a successor paradigm to PUS. While PUS has been predicated on there being a deficit in the 'public understanding of science', 'science and society' has focused on scientists' failures to adequately communicate science to general publics.

2 In March 2016, the state school board of Alabama voted unanimously to include a one-page disclaimer that included the sentence: 'The theory of evolution by natural selection is a controversial theory that is included in this textbook' (Goldston and Kyzer, 2009: 790).

3 Hasan's (2007) data on evolution were published by Hameed (2008) in the journal *Science*. It has been cited 34 times in academic papers (Web of Science), and is listed as in the 97th percentile of 'high attention' science papers of the same longevity.

4 This survey was used as part of the wider 'Darwin Now' project run by Professor Fern Elsdon-Baker. This was a project that sought to celebrate the duel anniversaries in 2009 of both Charles Darwin's birth and the publication of *On the origin of species*. This project ran in up to 50 countries worldwide and included a flagship conference in November 2009 at the Bibliotheca Alexandrina Egypt – hence the inclusion of Egypt in this survey.

5 In response to the question: 'Listed below are a range of different views, which people may or may not have about the origins of species and development of life on earth. Which of these comes closest to your own view?'.

6 It is important to note that the terms 'micro-evolution' and its counterpart 'macro-evolution' are products of creationist discourse rather than emerging from evolutionary science.

7 Falade and Bauer (2018), in their survey of the impact of religious identity on general attitudes to science among the general public in Lagos, Nigeria, offer a slightly modified version of this: (1) cognitive *polyphasia* of scientific and religious knowing manifesting as a 'hierarchy', when one form is elevated over the other; (2) 'parallelity', when both serve separate functions; and (3) 'empowerment', where one enhances the other.

8 Recent events in Turkey illustrate how curricula are subject to political change. After the 1917 republican revolution, Ataturk's government pursued a policy of promoting the acceptance of evolution as in line with modern, secular values (Tee, 2016). Debates over evolution have consequently mapped on to secular–Islamist debates in Turkey. Since the failure of the 2016 coup and the consequent strengthening of the Erdogan presidency, the ruling AKP party has considered removing evolution from the school curriculum. While the AKP government may be emboldened to alter education policy to exclude evolution due to its current unassailable political dominance, the executive of the Moroccan government (effectively the inner circle dominated by the 'modernizing' King Mohammed VI) seems recently to have watered down evolutionary science in the school curriculum in deference to Morocco's Higher Council of Religious Scholars. This may be a move to bolster the King's religious credentials in response to criticism of the monarchy's power from Islamist political opponents.

## References


Aroua, S., Coquide, M. and Abbes, S. (2009) 'Overcoming the effect of socio-cultural context: impact of teaching evolution in Tunisia', *Evolution: Education and Outreach*, 2(3): 474–8.

Asghar, A. (2013) 'Canadian and Pakistani Muslim teachers' perceptions of evolutionary science and evolution education', *Evolution: Education and Outreach*, 6(10): 1–12.

Asghar, A., Wiles, J.R. and Alters, B. (2007) 'Discovering international perspectives on biological evolution across religions and cultures: insights gained through developing methodological tools for research in diverse contexts', *International Journal of Diversity*, 7: 81–8.

Asghar, A., Hameed, S. and Farhani, N. (2014) 'Evolution in biology textbooks: a comparative analysis of 5 Muslim countries', *Religion and Education*, 41(1): 1–15.

Barbour, I.G. (1990) *Religion in an age of science: The Gifford lectures, volume one*, San Francisco, CA: HarperCollins.

Bauer, M.W., Allum, N. and Miller, S. (2007) 'What can we learn from 25 years of PUS survey research? Liberating and expanding the agenda', *Public Understanding of Science*, 16(1): 79–95.

Berkman, M. and Plutzer, P. (2010) *Evolution, creationism, and the battle to control America's classrooms*, New York, NY: Cambridge University Press.

Bhattacharjee, Y. (2010) 'NSF board draws flak for dropping evolution from indicators', *Science*, 328(5975): 150–1.

Bodmer, W. (1985) *The public understanding of science*, London: Royal Society.

BouJaoude, S., Asghar, A., Wiles, J.R., Jaber, L., Sarieddine, D. and Alters, B. (2011a) 'Biology professors' and teachers' positions regarding biological evolution and evolution education in a Middle Eastern society', *International Journal of Science Education*, 33(7): 979–1000.

BouJaoude, S., Wiles, J.R., Asghar, A. and Alters, B. (2011b) 'Muslim Egyptian and Lebanese students' conceptions of biological evolution', *Science and Education*, 20(9): 895–915.

Burton, E.K. (2011) 'Evolution and creationism in Middle Eastern education: a new perspective', *Evolution*, 65(1): 301–4.

Chan Kit Yok, M., Clement, P., Leong, L.K., Lee Shing, C. and Anak Ragem, P. (2015) 'Preliminary results on Malaysian teachers' conception of evolution', *Procedia – Social and Behavioural Sciences*, 167: 250–5.

Clark, M. (2009) *Islam and new kinship: Reproductive technology and the Shariah in Lebanon*, New York, NY: Berghan Books.

Clément, P. (2015) 'Muslim teachers' conceptions of evolution in several countries', *Public Understanding of Science*, 24(4): 400–21.

Coyne, J.A. (2012) 'Science, religion and society: the problem of evolution in America', *Evolution*, 66(8): 2654–63.

Croché, S. (2015) 'Science and religion on the blackboard: exploring schoolmasters' beliefs and practices in Senegal', *British Journal of Religious Education*, 37(1): 37–52.

Dagher, Z.R. and BouJaoude, S. (1997) 'Scientific views and religious beliefs of college students: the case of biological evolution', *Journal of Research in Science Teaching*, 34(5): 429–45.

Deniz, H, Çetin, F. and Yilmaz, I (2011) 'Examining the relationships among acceptance of evolution, religiosity, and teaching preference for evolution in Turkish preservice teachers', *Reports of the National Center for Science Education*, 31.4, 1.1–1.9.

Elsdon-Baker, F. (2015) 'Creating creationists: the influence of "issues framing" on our understanding of public perceptions of clash narratives between evolutionary science and belief', *Public Understanding of Science*, 24(4): 422–39.

Elsdon-Baker, F. (2018) 'Re-examining "creationist" monsters in the uncharted waters of the social study of science and religion', in B. Nerlich, S. Hartley, S. Raman and A. Smith (eds) *Science and the politics of openness: Here be monsters*, Manchester: Manchester University Press.

Elsdon-Baker, F., Leicht, C., Mason-Wilkes, W., Preece, E. and Piggott, L. (2017) 'Summary report of preliminary findings for a survey of public perspectives on evolution and the relationship between evolutionary science and religion'. Available at: https://sciencereligionspectrum.org/wp-content/uploads/2017/09/SRESYouGov-survey-preliminary-findings-5.9.17.pdf (accessed 14 April 2018).

Evans, M.S. and Evans, J.H. (2010) 'Arguing against Darwinism: religion, science, and public morality', in B.S. Turner (eds) *The new Blackwell companion to the sociology of religion*, Hoboken, NJ: Blackwell Publishing Ltd, pp 286–308.

Eve, R.A. and Harrold, F.B. (1991) *The creationist movement in modern America*, Boston, MA: Twayne Publishers.

Everhart, D. and Hameed, S. (2013) 'Muslims and evolution: a study of Pakistani physicians in the US', *Evolution: Education and Outreach*, 6(1). Available at: https://evolution-outreach.springeropen.com/articles/10.1186/1936-6434-6-2 (accessed 16 April 2018).

Falade, B.A. and Bauer, M.W. (2018) '"I have faith in science and in God": common sense, cognitive polyphasia and attitudes to science in Nigeria', *Public Understanding of Science*, 27(1): 29–46.

Goldston, M.J. ('Dee') and Kyzer, P. (2009) 'Teaching evolution: narratives with a view from three southern biology teachers in the US', *Journal of Research in Science Teaching*, 46(7): 762–90.

Guhin, J. (2016) 'Why worry about evolution? Boundaries, practices, and moral salience in Sunni and evangelical high schools', *Sociological Theory*, 34(2): 151–74.

Guillo, D. (2017) 'Representations of the *Origin of species* in secular (France) and religious (Morocco) contexts', in K. Rutten, S. Blancke and R. Soetaert (eds) *Rhetorical, cognitive and epistemological perspectives on science and culture*, West Lafayette, IN: Purdue University Press, pp 171–92.

Hamdy, S. (2013) *Our bodies belong to God: Organ transplants, Islam, and the struggle for human dignity in Egypt*, Berkley, CA: University of California Press.

Hameed, S. (2015) 'Making sense of Islamic creationism in Europe', *Public Understanding of Science*, 24(4): 388–99.

Hameed, S. (2008) 'Bracing for Islamic creationism', *Science*, 322(5908): 1637–8.

Hassan, R. (2007) 'On being religious: patterns of religious commitment in Muslim societies', *The Muslim World*, 97(3): 437–76.

Hill, J.P. (2014) 'Rejecting evolution: the role of religion, education, and social networks', *Journal for the Scientific Study of Religion*, 53(3): 575–94.

Hrairi, S. and Coquidé, M. (2002) 'Attitudes d'élèves Tunisiens par rapport à l'évolution biologique', *Hétérogénéité et Différenciation*, 35: 149–63.

IAP (InterAcademy Panel on International Issues) (2006) 'Statement on the teaching of evolution', IAP. Available at: www.interacademies.org/13901/IAP-Statement-on-the-Teaching-of-Evolution (accessed 21 January 2019).

Inhorn, M.C. (2004) 'Middle Eastern masculinities in the age of new reproductive technologies: male infertility and stigma in Egypt and Lebanon', *Medical Anthropology Quarterly*, 18(2): 34–54.

Kahan, K.M. (2015) 'Climate science communications and the measurement problem', *Advances in Political Psychology*, 36: 1–43.

Larson, E.J. (2007) *The creation–evolution debate: Historical perspectives*, Athens, GA: University of Georgia Press.

Lienesch, M. (2007) *In the beginning: Fundamentalism, the Scopes trial, and the making of anti-evolution movement*, Chapel Hill, NC: University of North Carolina Press.

Live Science (2006) 'US lags world in grasp of genetics and acceptance of evolution', *Live Science*, August 10. Available at: www.livescience.com/963-lags-world-grasp-genetics-acceptance-evolution.html (accessed 21 January, 2019).

Mansour, N. (2010) 'Science teachers' views of science and religion vs. the Islamic perspective: conflicting or compatible?', *Science Education*, 95(2): 281–309.

Michigan State University (2006) 'Americans less likely to accept evolution than Europeans', *MSUToday*, 10 August. Available at: https://msutoday.msu.edu/news/2006/americans-less-likely-to-accept-evolution-than-europeans/ (accessed 11 October 2018).

Miller, J.D., Scott E.C. and Okamoto, S. (2006) 'Public acceptance of evolution', *Science*, 313,(5788): 765–6.

Moran, J.P. (2012) *American genesis: The evolution controversies from Scopes to creation science*, New York, NY: Oxford University Press.

Noy, S. and O'Brien, T.L. (2018) 'An intersectional analysis of perspectives on science and religion in the United States', *The Sociological Quarterly*, 59(1): 40–61.

Numbers, R.L. (2006) *The creationists: From scientific creationism to intelligent design*, Cambridge, MA: Harvard University Press.

Paz-y-Miño, C.G. and Espinosa, A. (2015) 'Evolution controversy: a phenomenon prompted by the incompatibility between science and religious beliefs', *The International Journal of Science in Society*, 7(2): 1–23.

Peker D.C., Comert, G.G. and Kence, A. (2010) 'Three decades of anti-evolution campaign and its results: Turkish undergraduates' acceptance and understanding of the biological evolution theory', *Science & Education*, 19, 739–55.

Pennock, R.T. (2000) *Tower of Babel: The evidence against the new creationism*, Cambridge, MA: MIT Press.

Pew Research Centre (2013) 'The world's Muslims: religion, politics and society', The Pew Forum on Religion and Public Life. Available at: http://assets.pewresearch.org/wp-content/uploads/sites/11/2013/04/worlds-muslims-religion-politics-society-full-report.pdf (accessed 3 October 2018).

Rosengren, K.S., Brem, S.K., Evans, E.M. and Sinatra, G.M. (2012) *Evolution challenges: Integrating research and practice in teaching and evolution*, New York, NY: Oxford University Press.

Ruse, M. (2005) *Evolution–creation struggle*, Cambridge, MA: Harvard University Press.

Shapiro, A.R. (2015) *Trying biology: The Scopes trial, textbooks, and the antievolution movement in American schools*, Chicago, IL: University of Chicago Press.

Stears, M., Clement, P., James, A. and Dempster, E. (2016) 'Creationist and evolutionist views of South African teachers with different religious affiliations', *South African Journal of Science*, 112(5/6). Available at: https://pdfs.semanticscholar.org/df8a/820163b0efdf9a9d4aae5273d90874375d2a.pdf (accessed 3 October 2018).

Suldovsky, B. (2016) 'In science communication, why does the idea of the public deficit always return? Exploring key influences', *Public Understanding of Science*, 25(4): 415–26.

Taşkın, O. (2014) 'An exploratory examination of Islamic values in science education: Islamization of science teaching and learning via constructionism', *Culture Studies of Science Education*, 9: 855–75.

Tayob, A. (2012) 'Politics and Islamization in African public spheres', *Islamic Africa*, 3(2): 139–62.

Tee, C. (2016) *The Gülen movement in Turkey: The politics of Islam and modernity*, London: I.B. Tauris.

Unsworth, A. and Voas, D. (2018) 'Attitudes to evolution among Christians, Muslims and the non-religious in Britain: differential effects of religious and educational factors', *Public Understanding of Science*, 27(1): 76–93.

Webb, G.E. (1994) *The evolution controversy in America*, Lexington, KY: University of Kentucky Press.

Wiles, J.R. (2011) 'Challenges to teaching evolution: what's a head?', *Futures*, 43: 787–96.

Yousuf, A., Daud, M.A. and Nadeem, A. (2011) 'Awareness and acceptance of evolution and evolutionary medicine among medical students in Pakistan', *Evolution: Education and Outreach*, 4(4): 580–8.

PART III

# Science, Culture and Non-religion

8

# Feeling Rational: Affinity and Affinity Narratives in British Science–Non-religion Relations

*Lois Lee*

## Introduction

The idea that religion and science are opposed to one another has proved remarkably resilient, despite widespread, wide-ranging and often quite straightforward and easily graspable criticisms of its core premises. Given that the idea of science and religion as being intrinsically conflicting now seems naive, the question to ask is not 'How are religion and science opposed to one another, if at all?', but rather 'What is the basis of this flawed idea of conflict and its hold over popular and scholarly imaginations?'. If the idea of a science–religion conflict is not based in reality, how and why does it persist as one of the central narratives about religion–science relations, and about modernity more generally?

The goal of this chapter is to consider this question from the perspective of non-religion and secularity studies (Bullivant and Lee, 2012), focusing, in particular, on the relationship between science and non-religiosity.[1] These areas of study offer an important vantage point since the other side of the coin to the idea of science–religion conflict is the idea of science–non-religion *affinity*. It is therefore possible and potentially helpful to ask what role the latter idea plays in sustaining the popularity of the former.

This discussion builds on ideas arising from critical secular studies and critical religion studies, both of which challenge the idea that science mainly impacts on religion epistemically, and instead draw attention to the ideological and mythological roles that science plays in the subjectivities, identities and cultures of non-religious people.[2] This work tends to be largely theoretical or to rely on very broad historical accounts of Western modernity in its entirety (Lee, 2019). The aim of this chapter is therefore to contribute to – and further encourage – the more localized and detailed empirical explorations of perceived non-religion–science affinity that are just beginning to emerge (see, eg, Kaden et al, 2017; see also Elsdon-Baker et al 2017).

The chapter uses the UK as a case study for understanding the role of science within non-religious cultural formations found more widely, especially across Europe and other Western regions. It returns to data from the 'Being Secular' research project (Lee, 2015), gathered primarily through in-depth interviews with people who said that they would prefer to identify as 'non-religious' than 'religious'. These interviews ($n$ = 42) were conducted in Cambridge and Greater London between 2006 and 2011, and explored participants' understandings and experiences of religious cultures and 'religious-like' things (eg life-cycle ceremonies) in relation to their non-religious identification.[3] For this chapter, I have reviewed these conversations in light of questions posed by the editors of this volume and the broader 'Science and Religion: Exploring the Spectrum' (SRES) project about science–religion relations. The chapter also draws on wider ethnographic material from the 'Being Secular' project, more recent media material relevant to the British context and a number of recent empirical studies from across the human sciences that provide further insights into the relationship between non-religion and science.

I argue that these explorations indicate that the perceived affinity between non-religion and science has some empirical grounding, and that this needs to be taken seriously. However, I also argue against naive understandings of this affinity and its implications. Rather, this work supports the view that the real force of these relations is cultural and historic, rather than intellectual and epistemic. This idea of a non-essential connection between non-religious and scientific cultures is consistent with critical accounts; however, the argument builds on these in highlighting the ambivalence, variability and heterogeneity of these relations. Critical approaches focus on how the non-religious caricature the religious as unscientific but neglect the more constructive role that science can also play in non-religious people's meaning systems; by contrast, this work seeks to highlight that non-religious people's engagement with science does not always refer to others or impact upon them, so much as it offers the non-religious person personal and inward-facing rewards, as a source of comfort and meaning as they go about their lives. Past scholarship has also tended to generalize the nature of science–non-religion relations, but this empirical work demonstrates that science does not play the same role in the lives of all non-religious people. Again, this observation is consistent with other data showing that non-religious populations are split in their attitudes towards particular scientific claims (eg about evolution) or science in general (Catto, 2017; Elsdon-Baker, 2017; see also Hill, Chapter 2, this volume). I argue that differentiating broad non-religious populations according to their different *existential outlooks* and *cultures* (Lee, 2015; see also Lee and Bullivant, 2016) can help to understand these differences. I propose that there are significant intellectual and cultural affinities between science and existential humanism in its rationalist form, but that these affinities are not seen in relation to other forms of humanism, nor into other non-religious existential cultures such as atheistic agnosticism, the latter of which involves considerable ambivalence towards science. I begin with a short exploration of the role and significance of the study of non-religious actors and cultures for understanding the relationship between religion and science.

## Science, non-religion and existential culture

Despite wide-ranging challenges to the religion–science conflict narrative, it remains both a motif in popular discourse in Western contexts, and a cornerstone of academic theories about the nature and role of religion and secularity in contemporary societies. As Kaden et al (2017: 1) put it, the 'science vs. religion' rubric has, over time, spawned its own industry, and its ability 'to excite public and professional passions' has hardly diminished. If this 'rubric' is not empirically grounded in any straightforward way, what, then, are its foundations? This chapter calls attention to an aspect of science–religion relations that is normally implicit but is nevertheless integral to it: the idea that people who say they have no religion – referred to in this chapter as 'the non-religious' – have a distinctively positive relationship with science.

In this view, the non-religious individual's relationship to science is, at the very least, unimpeded in the way in which a religious individual's is. However, others go much further. Critical scholarship of religion and secularity suggests, for example, that the idea of science–non-religion affinity has its own momentum – that actors are invested in this idea in its own right. Science–non-religion affinity is also identified as a narrative that both religious *and* non-religious actors are invested in (Fitzgerald, 2000). For example, the non-religious might enjoy claiming that they have privileged access to scientific knowledge and reason, but this idea can also sustain the claim that religious actors have privileged access to the realms of existential or 'spiritual' meaning, myth and even the arts. So, we find that religious authorities often distance themselves from 'the inquiring mind', spoken of with pity or contempt, and seen as a cause of irreligion (Lee, 2015: 57, 167). At the same time, non-religious voices often lay claim to the idea that their minds are more rational, or, more specifically, that their existential and metaphysical outlooks are more scientific.

Other work shows how these logics of science–non-religion affinity are matters of identity – of self-understanding and representation of the self in public settings (Catto, 2017). These ideas and patterns of identification are found in – and inculcated through – popular culture in the UK. This is most obvious perhaps in prominent television and radio science programming, which is frequently associated with figures who are known for and explicit about their non-religious philosophies. Recent examples include work by popular science broadcasters Brian Cox and Robin Ince, who follow 20th-century forebears in creating science programming that often interprets this material according to a humanist outlook.[4] However, the association between science and non-religion is found much more widely, and not only in programming explicitly associated with science. In 'Crazy ex-girlfriend' – a US musical comedy television drama series first screened in the UK in 2015, and focused on themes very little concerned with matters of science or philosophy – the opening scenes find the heroine, Rebecca, making the following speech in a moment of panic: 'Dear God; *I don't pray to you, because I believe in science*, but I don't know what to do. Give me guidance. Please. Ahh-men…. "Ay-men"?! "Ahh-men"?!' (emphasis added). These kinds of casual references found within mainstream programming illustrate the prevalence of 'belief in science' as a central way of understanding and identifying a non-religious outlook – and incidentally complicate notions of the US as normatively religious (Berger et al, 2008; Trzebiatowska, 2018).

However, as established as ideas about the close relationship between science and non-religion may be, they have not been closely examined from a social-scientific perspective until very recently. Although numerous scholars have challenged the notion of a simple opposition between science and religion (eg Fitzgerald, 2000; Asad, 2003) as well as a series of other binaries that often operate in intersection with it (the rational–irrational and the modern–pre-modern), this work has not always paid close empirical attention to exactly how science is entwined into the cultures and identities of the non-religious *qua* non-religious – of what science means to them in their everyday lives. This chapter aims to contribute to this work of understanding what exactly the relationship between the non-religious and science is.

In so doing, it draws on insights from recent empirical studies working with non-religious populations – normally operationalized as those identifying as non-religious or with another label that is interpreted as non-religious in the local context (eg 'atheism' or 'humanism' in the UK). This work is increasingly interested in the 'religious-like' beliefs and cultures of these individuals and populations (eg Lee, 2015; Baker and Smith, 2016; Taves, 2016), and in my own work, I am interested in how non-religious populations include not one, but several alternative 'world views', or what I refer to as *existential cultures* (Lee, 2015). The idea of existential culture is not seen as exclusive to non-religious people: many religious traditions are characterized by (though by no means reducible to) the way in which they offer conceptualizations of existence and reality too – but non-religious people conceptualize their existence in similar ways. These conceptualizations are intrinsically transcendent or metaphysical since they do not merely participate in secular, 'this-worldly' existence and reality, but make or otherwise manifest claims *about* the nature of that existence and reality (cf Simmel, 2010). Although it may be possible to think of religious traditions as forms of 'existential culture', the way in which non-religious individuals and groups are also observed to conceptualize the nature of existence and reality means that this notion offers a way of understanding non-religious populations more specifically, too.

In this chapter, I use this notion of existential culture to explore the way in which different non-religious existential orientations engage with science in distinctive ways. In so doing, I put to one side questions about the role of science in non-religious people's secular or 'this-worldly' lives. As Farias et al (2013: 1211, emphasis added) say, 'most individuals accept science as a reliable source of knowledge about *the world*'; this is true of people with religious as well as non-religious identities and meta-existential beliefs, especially in the UK (Elsdon-Baker et al, 2017). Rather, the chapter focuses on the several ways in which science is bound up with non-religious existential cultures that transcend the secular world through the processes and performances of making sense of it. This chapter also draws attention to non-religious existential outlooks and cultures that are more critical of or distant from science. Rather than the blanket and largely ideological scientism that critical approaches diagnose the non-religious to have, this investigation adds weight to the argument that there is a significant affinity between science and non-religion, which is historical and powerful but neither intrinsic nor universal.

## Science in non-religious existential life: rationalist humanism

The first part of this chapter focuses on a non-religious existential culture within which science plays a notable role, namely, rationalist humanism. Rationalist humanism is a materialist outlook that places emphasis on humanity and, in particular, humans' rational capacities as a source of existential meaning. When scholars and others identify a cultural affinity between the non-religious and science, it is most often the rationalist humanist that they have in mind. Indeed, one scholar (LeDrew, 2015) even suggests the term 'scientific atheism' in his work to describe that part of the US secular movement,[5] while the central role of science in the works of those most prominent of rationalist humanists, the New Atheists, is well known.

In fact, New Atheism provides a helpful example of rationalist humanism and its relationship with science. Although the term 'New Atheism' is contested, it tends to denote a cultural movement that emerged in the mid-2000s and is particularly associated with Richard Dawkins, Sam Harris, Daniel Dennett, and Christopher Hitchens, as well as later-comers such as A.C. Grayling.[6] For all their differences, these writers are united by two things: a radical secularist agenda that advocates the marginalization of religion, whether on the basis of Enlightenment rationalist claims about the disaffinity of religion and reason or of moral objections about religious authority and institutions; and veneration of the achievements and aesthetics of Western science. The role of science is central. In one scholar's words, 'New Atheism is *built* upon a rhetorical distinction between religion and scientific knowledge or, more specifically, a distinction between religion and science or religion and rationality'; it is 'a popular, media-driven discourse in which a strict distinction between religion and scientific knowledge is *a defining feature*, functioning as a means of drawing boundaries around the opposing parties involved in the debate' (Taira, 2012: 98, emphasis added). This iteration arguably overstates the role of the religion–science conflict thesis within New Atheism relative to its other concerns; several contributors are, for example, more concerned with their moral objections to religion. However, it captures the significant role that science plays within New Atheist culture, as well as critical scholars' account of this role, as a matter of identity and identity politics. In Taira's (2012: 98) words: 'Despite the New Atheist claims to represent universal scientific knowledge and rationality, their popularity is related to the public realm of identity politics, in which atheists demand recognition as atheists'.

The New Atheist case certainly provides further support for the view that science plays a role in the identities of rationalist humanists, and maybe other non-religious orientations besides. However, it is important to recognize that the rationalist humanist connection with science is much more diverse than New Atheism-focused critiques acknowledge. For one, science is also a genuine source of existential meaning and comfort for some – an effect also documented in social-psychological research (see Farias et al, 2013). This use of science as an existential resource was articulated in my fieldwork by Matthew,[7] a retired academic in his 70s. Matthew told me that "science is what is important to me". He explained:

> MATTHEW: "I believe in science as a frame of reference, I think that in ... the sense of finding one's place in the universe ... the method and the whole perspective of a scientific age is one that I find very comforting. I don't need supernatural explanations because ... nature in itself is impressive enough."
>
> INTERVIEWER: "Can you say more about that? What's comforting about it? Is it not something to...?"
>
> MATTHEW: "Well, you mentioned earlier that there's this notion of fear, yes, and comfort is somehow a response to fear and fear derives from uncertainty and ... the nature of the universe and the powers at large of the universe creates uncertainty, which can require comfort through some kind of intellectual system, which is based on the superstition or the supernatural, which I personally don't feel the need."
>
> INTERVIEWER: "So, science resolves – or means you're not uncertain and therefore... Is that...?"
>
> MATTHEW: "I'm not uncertain about the things that a lot of people seem to be uncertain about, I'm not worried about the afterlife, I'm not worried [about] divine retribution, you know, or divine reward for that matter. I think this ... all belongs in the past but if people don't

> want to embrace science and they feel comfort by traditional remedies, that's fine by me."

This is the rationalist orientation that critical secular and other scholars most associate with non-religious identification: it finds in scientific knowledge a source of existential comfort, and it dismisses religiosity as a source of comfort enabled by a disinclination to 'embrace science'. However, for Matthew, the most important thing is not identifying or distancing himself from a religious other; rather, those remarks are used to help him articulate his own engagement with science as a 'frame of reference' and a source of comfort. At the same time, the example illustrates how ideas of science–religion disaffinity are a resource that is drawn on to help articulate the central role of science within a non-religious existential philosophy.

There is other evidence that, outside of the New Atheism, this kind of 'belief in science' is found quite widely, and this work supports the idea that this belief is not only a matter of identification and identity politics. Farias et al (2013: 1210) argue, for example, that their social-psychological research shows that science can act as, 'a form of "faith" that helps [non-religious people] to deal with stressful and anxiety provoking situations'. This research differentiates this 'belief in science' from the more general and more widely held view that science provides legitimate knowledge about the material world since these 'believers' expand the remit of scientific knowledge to include existential and metaphysical questions and subsequently elevate its value as a source of meaning (Farias et al, 2013: 1211). This throws more light on how science–non-religion affinity narratives can translate into the idea of science–religion disaffinity since the kind of value that science plays within the *existential* lives of non-religious people may be incompatible with religious existential ideas. Although this affinity only relates to the existential domain and has no necessary bearing on how individuals relate to science as an aspect of secular life, it may be that this sense of affinity has the capacity to burst its banks – especially since, in practice, our lives are not lived accordingly to neatly demarcated domains (the economic, the political, the existential, etc), but rather finds them constantly and messily tangled together.

## Science in 'lived' rationalist humanism

Although Matthew casts science as an 'intellectual system', rationalist humanists also connect to science in non-intellectual ways. In the course of my fieldwork, I was struck by an informal conversation about my research with a young woman living in the area. I was talking with her about emerging evidence that non-religious outlooks and identities can arise through socialization processes, just as religious ones do, when she interjected – with some exasperation – 'But it *feels* rational!' (Lee, 2015: 188, emphasis in original). What struck me then was twofold: first, the force of feeling with which this was said; but then, second, the way in which her words expressed an emotional connection to the rational – her insistence was not that her position *was* rational, but that it *felt* rational. She was perhaps simply trying to express her sense that the argument I was making was not convincing but was not quite able to articulate why. However, her remark also speaks to the way in which the role of science and the rational can be experienced by non-religious people (as others) in ways that are not purely intellectual, but multilayered and embedded in their lived lives in more complicated ways. A purely critical account of rationalist humanism misses something when it does not capture these dimensions.

For example, Katie Aston's (2019) ethnographic research has explored the materiality of culture connected with British non-religious organizations such as Humanism UK (formerly the British Humanist Association) and the Rationalist Association – both organizations that have a strong rationalist history. Her work highlights, for example, the centrality of especially biological images in visual and material culture, including the pervasive use of images of, or relating to, Charles Darwin. These images manifest not as scientific references per se, but as icons around which people form and express their identities *as* people connected to a particular existential culture. Aston's work demonstrates the nuance that we need to bring to our understanding of these individuals and groups since, in practice, their boundary-making practices are, like all boundary making perhaps, ambivalent: they involve identity formation, community building and meaning making, as well as identity politics and the processes of othering involved in that.[8]

Rationalist humanism and the role in which it casts science may also be encountered in everyday life through representations in popular culture. On this, media historian Alexander Hall's (2017) idea of the 'humanist blockbuster' is instructive. 'Humanist blockbusters' are the expensive BBC science and nature programmes of the 20th century that make use of emotive language, sweeping scores and, in more recent examples, richly cinematic cinematography to inculcate an intensely *emotional* response to humanist narratives of existence. Although Richard Dawkins and his image have been used as a marker of atheism in British public life (Lee, 2017), Hall's work prompts us to consider whether it is less controversial figures who front television series – Brian Cox or even the beloved BBC nature documentary presenter Sir David Attenborough – who exemplify the tradition of 20th-century humanist blockbusters that Hall documents, and that tell us more about the prevalence and standing of existential humanism in the UK. Certainly, it demonstrates how the idea of science–rationalist humanist affinity may be a part of everyday life, and highlights again how those encountering it may engage with it not only intellectually, but also emotionally and as an aesthetic experience as much as an epistemic one.

Taken together, all of these examples demonstrate that what is at stake for rationalist humanists faced with the idea that religion and science are *not* in conflict is not so much the legitimacy of religious world views as it is the legitimacy of their own. Ideas about science–religion conflict are woven into rationalist humanists' existential meaning systems, and these meaning systems matter to them. These attachments are at least partly responsible for the way in which science–religion conflict narratives can excite emotional responses as well as intellectual ones: rationalist humanists *feel* rational; rationalist humanists are *comforted* by science. These attachments are the result of cultural forms – encouraged, for example, through encounters with 'humanist blockbusters' – but it also helps bring those forms into being. Those who *perceive* there to be an affinity between science and non-religion generate cultural forms – identities, images and objects – expressing this idea, and around which they coalesce with others who share this perception. In turn, these cultural formations influence the world around them, including popular representations and understandings of science.[9] In these ways, the perception and subjective experience of science–non-religion affinity are realities that lend a particular kind of truth to the claim that this affinity really exists. It might be cultural rather than epistemic or epistemological, but this affinity is found within sociocultural contexts, and impacts upon them, just the same.

## A problem: disaffinities between rationalist humanism and science

So far, so friendly. Although we might need to nuance rationalist accounts of why science and non-religious existential cultures connect and cohere in certain ways, we cannot deny that they do so. It is important, however, not to overstate this affinity.

In reviewing interview transcripts from 'Being Secular' for this chapter, for example, I noted how infrequently the idea of science came up in interviews. It was far from absent – as the examples in this chapter demonstrate. However, most interviewees made reference to science only once, and sometimes not at all.

When they did so, science was most consistently mentioned in response to a question concerning participants' thoughts about the origins of the universe. The following are typical answers in response to the question 'What are your thoughts about how the universe came into being?':

> "Well, I'm into astronomy and I believe the Big Bang and, you know, evolution, the whole picture as modern-day contemporary science sets out that it does." (David, 57)

> "Erm, I think that science has told us a lot about that, I think that we don't know everything yet, obviously, but – and I don't think we ever will know everything, I think we'll keep learning. I think, you know, that Big Bang theory makes as much sense as it can do without fully understanding quantum physics and [laughter], you know, without being clever enough to know what all that means but I would go with the Big Bang theory of change over time in terms of evolutionary processes, yeah." (Cat, 26)

Interviewees with a less materialist orientation sometimes raised scientific explanations at this point, before going on to describe their limitations, but the association between science and questions of creation was widespread. This demonstrates not only that science may not be associated with non-religious (and other) existential philosophies in general, but also that it is invoked in relation to particular topics or themes – and considered less relevant to others.

It is striking, too, that most participants appeared to be particularly uninterested in this question of origins compared to other questions and topics we discussed. If science was considered particularly salient to this topic, the topic itself was not treated as particularly salient to the broader themes of our interview. One participant, Dora, a 78-year-old woman from London, put this feeling into words:

> INTERVIEWER: "What are your thoughts about how the universe came into being?"

> Dora: "I think we don't know. And I don't think it matters. Because what matters is how you live your life day to day.... these people who think God made the universe ... they've got a hard decision to make to think God made the universe in seven days, therefore I'll do so and such. I think it's totally irrelevant to what is important, which is how we live here today and how we treat other people.... I don't know whether science will ever give us.... As an academic, it's quite interesting, but it's totally irrelevant to my life and the way I live it."

Thus, issues bound up with science were often characterized as less relevant to existential beliefs and even identities.

Exploring what it means to be non-religious as a lived reality also draws attention to the fuzzy edges of rationalist humanists' engagement with science, and its openness to change. Carl (29, London) characterized himself as a person for whom "religion has zero or close to zero influence, whereas science, maybe 90 or something like that". He said that he was "an advocate for science", and he understood his views on religion and science as part of a zero-sum game. However, elsewhere, Carl took a more qualified stance. Discussing his friendship with Lisa, a friend of his with "this really strong Christian belief that only comes out in some situations", he said:

> "I don't think [her religion has] ever really been much of an issue other than broadened my mind. Before I met Lisa – I don't think it is connected, but around the same time – before, I would say, I thought science would explain everything that needs explaining. And meeting more people that had religious beliefs who weren't complete nutcases, I would say it's broadened my mind."

Carl clearly maintains a strong and rationalist idea of the religious other – as irrational, as a "nutcase" – but he shows how his own relationship with science is mutable.

Whereas critical approaches tend to talk in very general terms about how the idea of the non-religious as rational and more scientific has been naturalized, lived approaches point to the ways in which these ideas and discourses are sometimes situational and often qualified in certain ways. This raises problems for overly neat ideas about the relationship between science and non-religion: while we should take this relationship seriously as an empirical reality and one with positive as well as negative effects, and should not dismiss the idea of science–non-religion affinity as mere ideology, we should also recognize that this relationship is complex.

## A bigger problem: agnosticism and science

Building on these concerns about the nature and significance of the role of science within rational humanism, this next section calls attention to other non-religious existential orientations that do not share the same warm view of science in the first place. Indeed, there is even some evidence to suggest that, in some parts of the world, the UK included, a more ambivalent, even hostile engagement with science *as an existential matter* is more widespread within non-religious populations than the 'scientistic' modes of engagement associated with rationalist humanism. Recognizing these experiences may therefore require us to refine our sense of affinity between non-religion and science further still.

One of the most significant examples of a non-religious mode of existentiality that displays this more negative relationship with science is an existential culture that I refer to as agnosticism (Lee, 2015). Social scientists frequently make a distinction between those who reject theist claims outright and those who take the view that humans cannot know anything for certain about the existence of God/gods – conventionally (though problematically) referred to as 'atheists' and 'agnostics', respectively (eg Clements and Gries, 2017). One reason that this traditional nomenclature is problematic is that both groups largely fall within the 'negative' category of atheism in the sense that both atheists and agnostics 'live their lives as though God does not exist' (Brown, 2017: 439).[10] One indication of this is that a majority of agnostics identify as non-religious – though they are slightly more likely to identify with a religion than positive atheists are (see the 2008 British Social Attitudes survey, at: www.britsocat.com).[11] Here, however, I use the term 'agnosticism' to describe a broader existential orientation, one with some resemblance to postmodern and post-humanist orientations.[12] Drawing on the 'strong' rather than the 'weak' sense of the term, this existential agnosticism describes an orientation that centres on the limits of human existential and metaphysical knowledge (Lee, 2015). This involves not only the idea that humans cannot know anything of the existence of God, but also the idea that this lack of knowledge is indicative of a much broader state of existential ignorance.[13] For agnostics, the inaccessibility of existential knowledge informs particular ethical goods, aesthetics and mythologies, each of which are able to act as a source of existential comfort.

As a system for generating a particular type of knowledge, science is of limited existential value within agnosticism (though it may have great value for dealing with secular, 'this-worldly' concerns). While they may 'live their lives as though God does not exist', then, existential agnostics differ from rationalist humanists and other 'negative atheists' in their relationship with science. This alternative relationship with science is illustrated by Jude, a 31-year-old graduate student from Cambridge, who positions her outlook in contradistinction not only to religion, but also to science. Jude told me:

> I think that at times there's a sense of knowing within myself that it's something that doesn't require scientific or rational explanation, and it works on its own regardless.... I think over-analyzing that space kills it; I think belonging to religion kills it; because I prefer the idea that it's unexplained – because that makes life more exciting.... What I'm trying to say is that it's why I wouldn't be religious then, because it's the same thing as what science offers me: it offers me an end explanation. I'm not sure whether it's necessary to require an end explanation. (Lee 2015: 178)

This emphasis on the ineffable is typical of what I term the agnostic mode of existentiality. Agnostics are not frightened of the unknown (as some psychologists propose that humans universally are); they revel in it. They are, in a sense, much more frightened at the prospect of knowing too much because this detracts from some of what makes living meaningful and magical to them – the idea that it is unexplained.

This scepticism about the value of definitive knowledge about the nature and meaning of life devalues science to the extent that it is regarded as a system of knowledge – a way of providing answers. For some agnostics I interviewed, the existential is better engaged with through experiential modes of 'knowing' or being in the world, or through the arts, which can articulate these experiences in new, often beautiful ways.

However, we should be careful not to replicate what C.P. Snow famously called the 'two cultures' of arts versus science in understanding the differences between non-religious existentialities. In an interview with Tom, a young male writer from London who identified as an atheist, I encountered a rationalist account of the role of the arts as a source of existential explanation. Tom told me:

> "I mean, obviously, as a writer … my job is to sort of what Heaney calls raiding the inarticulate, and that's the sort of transcendental, the sublime or whatever. And that 'otherness' – like you're trying to articulate something like love or trying to articulate the feeling of loneliness and all these things are kind of traditionally the things that religion has, in the past, whether through ritual or through text, tried to sort of – these are the things that it's tried to explore and explain. And also provide answers for. And I suppose [writing] does the same thing….
>
> I mean it's amazing, sometimes you see things and you are literally speechless and that's because words can't describe it. And art is a good example of this; I think … art happens when the material in which something is expressed is so integral to the expression that it couldn't be articulated in any other way; it couldn't be paraphrased. And that seems to me to be related to God because … there's moments of living which feel like art. I suppose where some idea from the natural world or the world around us hails us in a way that we don't have words for. And artists tend to go about trying to close that gap."

For Tom, art is a distinctive form of knowing that can 'explain' and 'provide answers' that no other source of knowledge can; in the face of the unknown or unarticulated, it offers a unique way of 'closing the gap'. Tom's words offer a caution against assuming that just because science can be integral to rationalist humanism, it is necessarily so, and certainly not to the exclusion of all else. However, they also illustrate how existential outlooks can imbue different kinds of human action with different kinds of meaning. Although both Tom and Jude construe religion as seeking to offer existential explanation, for Tom, there is existential value in explanation in general; for Jude, there is not. These cases show that we cannot map existential values directly onto aesthetic ones, since – to put it in Farias et al's (2013) terms – 'belief in explanation' does not necessarily imply 'belief in science'.

## The scale of the problem

Compared to both atheist identification and rationalist humanism, agnosticism – as identity and as existential philosophy and culture – has received infinitesimal attention, both in social science research and in popular discourse. This does not mean, however, that it is absent from either. In fact, there are several indications that existential agnosticism is significant in British life.

In popular culture, the idea that meaning emerges from experiencing the unknown or unknowable world is a familiar trope, and is associated with major cultural movements such as Romanticism. In other work, I have discussed how David Attenborough, the nature broadcaster and one of the most popular and beloved figures in British public life, expresses something inclining towards the agnostic when he speaks of his joy in moving among the 'unrevealed complexity' of the natural world – an idea that is situated in contrast with Richard Dawkins's view that it is the excitement of a 'puzzle being solved' that gives meaning to his work (discussed in Lee, 2015: 176–7). Alongside those 'humanist blockbusters' (Hall, 2017), then, we might plausibly consider the role that agnosticism plays in the mainstream of British culture – though whether there are equivalent 'agnostic blockbusters' is an open and empirical question. Agnostic forms of meaning are certainly a common trope in the arts. The title poem from the prize-winning collection *Dear boy*, by Emily Berry, for example, laments:

> You rang me three times and said 'I can explain everything!'
> into my voicemail. You know perfectly well I believe
> nothing worthwhile is explainable. Dear boy. (Berry, 2013: 7)

In the social sciences, too, affinities between existential agnosticism and postmodernist theory, as well as with post-humanist philosophy, indicate a significant presence within the academy.

There are also some indications that existential agnosticism is widespread. For example, substantial numbers of people respond to survey questions about their belief in God with the strong agnostic answer – 'I don't know whether there is a God and I don't believe there is any way to find out'. Indeed, in the UK, the number of people selecting this option makes agnostics the single largest 'faith' category in the UK: 18%, compared to 17% who chose the 'positive atheist' statement in the last round of the British Social Attitudes survey to explore religion in depth (see the 2008 British Social Attitudes survey, at: www.britsocat.com). There is some evidence to back up the idea that these survey distinctions tell us something meaningful about the outlooks of those choosing them. These survey categories are, for example, bound up with different kinds of identity positions and politics: those who have positive atheist beliefs and those who have strong agnostic beliefs are different along a number of lines (Clements and Gries, 2017). Differentiating positive atheists and agnostics also complicates many of the established demographic correlations with atheism and non-religion. For example, both groups are famously more male than the theistic and the religious (eg Trzebiatowska and Bruce, 2012; Keysar and Navarro-Rivera, 2013) but it turns out that this is more true of positive atheists than it is of agnostics in the UK (see the 2008 British Social Attitudes survey, at: www.britsocat.com). These differences lend support to the idea that survey measures of positive atheism and strong agnosticism give some indication of the scale of existential agnosticism, as well as the idea that non-religious populations can be disaggregated into smaller populations with distinctive cultural orientations.

Given all this, the lack of empirical research into existential agnosticism is striking – and the need for new work is clear. There are significant practical reasons why agnosticism is not more widely engaged with in scholarship. There are, for example, great challenges in trying to sample and recruit from a population that does not tend to have clear representations or even self-understandings of its own existentiality – a difficulty shared by researchers of the 'indifferent to religion' population (Quack and Schuh, 2017). As with religious indifference, though, the indications we do have suggest that researchers need to take these challenges on if they are to truly understand the nature of our contemporary existential and religious landscapes.

## Conclusion

Focusing on the UK but placing these experiences in wider context, this chapter draws attention to several different ways in which science is – and is not – bound up with non-religious perspectives. This chapter has aimed to consolidate and deepen empirically grounded understandings of the role that science plays in the lives of non-religious people and, more broadly, to contribute to work seeking to understand why it is that religion–science conflict narratives are so resilient. We know that science can be bound up with non-religious people's self-understandings and identities (eg Elsdon-Baker, 2015), but it has not been clear (1) how science is incorporated in practice, and (2) how and why science does not *always* play this role within non-religious self-understandings and identities. The material discussed here shows that science, like religion and the arts, is bound up with the way in which people experience and try to articulate their existentialities, regardless of whether it is embraced or 'othered' in these processes. However, it also shows that, while science may play a significant role within some non-religious outlooks, this role is not consistent across all of them.

Specifically, this chapter proposes engaging with the existential beliefs and cultures (or 'world views') of non-religious people through empirical research in order to understand and theorize these discrepancies within non-religious populations. More and more, research working with non-religious populations lays emphasis on the existentiality of so-called 'unbelievers' (eg Droogers and Van Harskamp, 2014; Lee, 2015, 2019; Baker and Smith, 2015; Evans, 2016; Lagerkvist, 2013) – their beliefs about the nature and meaning of life, the social connections and ritual practices that partly constitute these beliefs, and so on (Lee, 2015). This chapter aims to show that the identification of these modes of 'existentiality' also provides a new and significant site for empirical investigations of the role of science – epistemic, mythic, ideological or otherwise. These modes of existentiality allow us to make meaningful distinctions within the general non-religious population and, through that, to build a more nuanced picture of non-religion–science relations. For example, paying close attention to – and undertaking empirical study of – the diverse beliefs and commitments of non-religious people helps us to understand why it is that, although non-religious people are more likely to perceive a conflict between some areas of science (eg evolutionary theory) and religion, a majority of them do not perceive this conflict (Elsdon-Baker, 2017). The contrasting role of science across non-religious existential cultures helps to understand these differences. Further empirical studies are needed, especially inductive work and research concerned with non-religious people's everyday lives – the sort of sociological 'lived' approaches that have been undertaken by Zuckerman (2008), Catto and Eccles (2013), Engelke (2015), Mumford (2015), myself and a handful of others.

In fact, it may well be that the distinctions made here between rationalist humanism and agnosticism in their relationship with science only skim the surface. Making this distinction is to say nothing of non-rationalist modes of existential humanism documented by Aston (2019) and LeDrew (2015), for example,[14] nor of the possibly many other forms of non-religious existentiality: the subjectivist cultures of the alternatively spiritual (Heelas and Woodhead, 2005); the non-human-centred forms of materialism, such as existential environmentalisms (sometimes referred to as 'deep ecology'); and perhaps many more besides. Each of these orientations also engages with science in distinctive ways, and if the relationships between the existential cultures examined here and science are multiple, it is likely that each of these unexamined orientations would also exhibit this same complexity.

By this light, we also need to urgently consider why it is that both rationalist humanism and positive atheism have received such disproportionate attention in academic research compared to other non-religious existential modes. This may be bound up with the idea that non-religion is synonymous with rationalist humanism, and that rationalist humanism's (distinctive) affinities with science therefore tell us everything we need to know about non-religion–science relations. Rather than accepting the oft-made claim that all non-religious people are scientistic and anti-religious, critical scholarship should maybe be asking other questions. How is it, for example, that rationalist humanists are so visible in public life, able to attract wide media coverage and shape public discourses much more than other non-religious existential cultures are able to do? Why have non-religion and secular studies focused so disproportionately on rationalist humanists (often identified by the term 'atheist') and hardly engaged at all with agnostics, as well as with other humanist outlooks? What are the structural forces channelling our scholarly imaginations, as well as wider public discourses, in these ways? Given that caricatures of the non-religious – as scientistic, among other things – are as common as caricatures of the religious as 'unscientific', these questions may be matters of social and political, as well as scholarly, concern.

## Acknowledgements

Many thanks to the editors of this volume for their help in developing this chapter. I am also grateful for the opportunity I had to develop this work at a workshop and conference convened by the 'Science and Religion: Exploring the Spectrum' (SRES) programme; thank you to the organizers and participants at those events. The chapter was also made possible through the support of grants from the John Templeton Foundation (grant numbers 59544 and 60624); however, the opinions expressed in this publication are those of the author and do not necessarily reflect the views of the John Templeton Foundation.

## Notes

1 I use 'non-religion' in the relational sense developed by Lee (2012, 2015), Quack (2014) and Cotter (2016) to identify phenomena that are understood (emically and/or etically) in contradistinction to religion. This usage of 'non-religion' is similar to the idea of the 'secular' in some of the senses it is used. However, I have argued that the term 'secular' is often confusing since it is also used to identify phenomena that are not necessarily non-religious and may even have significant religious aspects, such as 'secular states' that adopt a particular form of pluralist secularism in their constitution.

2 Lee (2019) provides a critical overview of these 'schools'.

3 In order to explore the potential variety of positions and meanings that underlie generic non-religious identifications, the sampling approach sought to maximize variation by working with people who differed according to an array of demographic characteristics (age, gender, race, religious background), although recruitment methods meant that the final sample was biased towards those with a tertiary-level education. A full review of the methodology used to gather these data can be found in Lee (2015).

4 Brian Cox is a popular science broadcaster who has been described as a 'humanist scientist' (Engelke, 2015), and Robin Ince is a comedian and patron of Humanists UK. On the BBC's 20th-century 'humanist blockbusters', see Hall (2017).

5 'Scientific atheism' is differentiated from 'humanistic atheism', though both orientations conform to the broader notion of humanism established in historical accounts. For example, LeDrew's 'scientific atheism' and 'humanistic atheism' would both be forms of what Charles Taylor (2007) calls 'exclusive humanism' in his seminal history of Christendom.

6 For a definition of New Atheism, see Lee and Bullivant (2016). For a wider discussion of New Atheism, see Cotter et al (2017).

7 All names have been changed to preserve anonymity.

8 Other research with non-religious movements outside of the UK has shown similar findings. See, for example, Smith and Cimino's (2015) US research and Kind's (Chapter 9, this volume) research with the Swedish Humanist Association.

9 We might consider whether the central role that Charles Darwin occupies in how modern science is imagined and represented in the UK is shaped, in part, by the significance of his work to prevalent forms of existential culture, including rationalist humanism, so that the *existential* salience of his work becomes a driver of his status within the scientific field and our conceptions of what science *is*. (My thanks to Stephen Jones for raising this point.)

10 My gratitude to Abby Day for bringing this very helpful definition of 'atheism' to my attention. For dictionary definitions of common understandings of concepts such as 'atheism' and 'agnosticism', and the difference between 'negative' and 'positive' atheisms, see Lee and Bullivant (2016).

11 The fact that not all non-theists or agnostics identify as non-religious is a moot point since religious identity and belief consistently overlap imperfectly; the same is true for those who identify as religious but do not believe in God.

12 On post-humanism, see Ferrando (2013).

13 The 'strong' sense of agnosticism indicates the view that nothing can be known about the existence of God; the 'weak' sense is used to describe a perspective of doubt and uncertainty – the proverbial fence-sitter (Lee and Bullivant, 2016).

14 LeDrew's notion of 'humanistic atheism' is an example of this, while his 'scientific atheists' would be described as 'rationalist humanists' within the conceptual framework used here.

## References

Asad, T. (2003) *Formations of the secular: Christianity, Islam, modernity*, Stanford, CA: Stanford University Press.

Aston, K. (2019) 'Formations of a secular wedding', in M. Scheer, N. Fadil and B. Shepelern Johansen (eds) *Secular bodies, affects, and emotions*, London: Bloomsbury Academic.

Baker, J.O. and Smith, B.G. (2015) *American secularism: Cultural contours of nonreligious belief systems*, New York, NY: New York University Press.

Berger, P.L., Davie, G. and Fokas, E. (2008) *Religious America, secular Europe? A theme and variations*, Aldershot: Ashgate.

Berry, E. (2013) *Dear boy*, London: Faber and Faber.

Brown, C.G. (2017) 'The necessity of atheism: making sense of secularisation', *Journal of Religious History*, 41(4): 439–56.

Bullivant, S. and Lee, L. (2012) 'Interdisciplinary studies of nonreligion and secularity: the state of the union', *Journal of Contemporary Religion*, 27(1): 19–27.

Catto, R. (2017) 'Science and religion conflict for non-religious Britons and Canadians', Nonreligion and Secularity Research Network. Available at: https://nsrn.net/2017/04/06/research-science-and-religion-conflict-for-non-religious-britons-and-canadians/ (accessed 27 April 2018).

Catto, R. and Eccles, J. (2013) '(Dis)Believing and belonging: investigating the narratives of young British atheists', *Temenos – Nordic Journal of Comparative Religion*, 49(1): 37–63.

Clements, B. and Gries, P. (2017) '"Religious nones" in the United Kingdom: how atheists and agnostics think about religion and politics', *Politics and Religion*, 10(1): 161–85.

Cotter, C.R. (2016) 'Religion-related discourse: a critical approach to non-religion in Edinburgh's Southside', PhD thesis, Lancaster University, UK.

Cotter, C.R., Quadrio, P.A. and Tuckett, J. (eds) (2017) *New atheism: Critical perspectives and contemporary debates*, Dordrecht: Springer.

Droogers, A. and Van Harskamp, A. (eds) (2014) *Methods for the study of religious change: From religious studies to worldview studies*, Sheffield: Equinox.

Elsdon-Baker, F. (2015) 'Creating creationists: the influence of "issues framing" on our understanding of public perceptions of clash narratives between evolutionary science and belief', *Public Understanding of Science*, 24(4): 422–39.

Elsdon-Baker, F. (2017) 'Science and religion: exploring the spectrum', keynote presentation at 'Science and Religion: Exploring the Spectrum', June, University of Manchester, UK.

Elsdon-Baker, F., Leicht, C., Mason-Wilkes, W., Preece, E. and Piggott, L. (2017) 'Summary report of preliminary findings for a survey of public perspectives on evolution and the relationship between evolutionary science and religion'. Available at: https://sciencereligionspectrum.org/wp-content/uploads/2017/09/SRESYouGov-survey-preliminary-findings-5.9.17.pdf (accessed 14 April 2018).

Engelke, M. (2015) '"Good without God": happiness and pleasure among the humanists', *Hau: Journal of Ethnographic Theory*, 5(3): 69–91.

Evans, J.H. (2016) *What is a human? What the answers mean for human rights*, Oxford and New York, NY: Oxford University Press.

Farias, M., Newheiser, A.K., Kahane, G. and De Toledo, Z. (2013) 'Scientific faith: belief in science increases in the face of stress and existential anxiety', *Journal of Experimental Social Psychology*, 49(6): 1210–13.

Ferrando, F. (2013) 'Posthumanism, transhumanism, antihumanism, metahumanism and new materialisms: differences and relations', *Existenz*, 8(2): 26–32.

Fitzgerald, T. (2000) *The ideology of religious studies*, Oxford: Oxford University Press.

Hall, A. (2017) 'Evolution on the small-screen: reflections on media, science and religion in twentieth-century Britain', conference presentation at 'Science and Religion: Exploring the Spectrum', June, University of Manchester, UK.

Heelas, P. and Woodhead, L. (2005) *The spiritual revolution: Why religion is giving way to spirituality*, Oxford: Blackwell.

Kaden, T., Jones, S.H., Catto, R. and Elsdon-Baker, F. (2017) 'Knowledge as explanandum: disentangling lay and professional perspectives on science and religion, *Studies in Religion/Sciences Religieuses*, 47(4): 1–22, DOI: 10.1177/0008429817741448.

Keysar, A. and Navarro-Rivera, J. (2013) 'A world of atheism: global demographics', in S. Bullivant and M. Ruse (eds) *The Oxford handbook of atheism*, Oxford: Oxford University Press, pp 553–86.

Lagerkvist, A. (2013) 'New memory cultures and death: existential security in the digital memory ecology', *Thanatos*, 2(2): 1–17.

LeDrew, S. (2015) *The evolution of atheism: The politics of a modern movement*, New York, NY: Oxford University Press.

Lee, L. (2012) 'Talking about a revolution: terminology for the new field of nonreligion studies', *Journal of Contemporary Religion*, 27(1): 129–39.

Lee, L. (2015) *Recognizing the non-religious: Reimaging the secular*, Oxford: Oxford University Press.

Lee, L. (2017) 'Vehicles of New Atheism: the atheist bus campaign, nonreligious representations and material culture', in C.R. Cotter, P.A. Quadrio and J. Tuckett (eds) *New Atheism: Critical perspectives and contemporary debates*, Dordrecht: Springer, pp 69–86.

Lee, L. (2019) 'Observing the atheist at worship: ways of seeing the secular body', in M. Scheer, N. Fadil and B. Shepelern Johansen (eds) *Secular bodies, affects, and emotions*, London: Bloomsbury Academic.

Lee, L. and Bullivant, S. (2016) *Dictionary of atheism*, Oxford: Oxford University Press.

Mumford, L. (2015) 'Living non-religious identity in London', in L.G. Beaman and S. Tomlins (eds) *Atheist identities: Spaces and social contexts*, New York, NY: Springer, pp 153–70.

Quack, J. (2014) 'Outline of a relational approach to "nonreligion"', *Method and Theory in the Study of Religion*, 26(4/5): 439–69.

Quack, J. and Schuh, C. (eds) (2017) *Religious indifference: Themes and conceptualization*, Dortrecht: Springer.

Simmel, G. (2010) 'Life as transcendence', *The view of life: Four metaphysical essays with journal aphorisms*, translated by John A.Y. Andrews and Donald N. Levine. Chicago, IL: Chicago University Press.

Smith, C. and Cimino, R. (2015) *Atheist awakening: Secular activism and community in America*, New York, NY: Oxford University Press.

Taira, T. (2012) 'New Atheism as identity politics', in M. Guest and E. Arweck (eds) *Religion and knowledge: Sociological perspectives*, Farnham: Ashgate, pp 97–114.

Taves, A. (2016) 'On the virtues of a meaning systems framework for studying nonreligious and religious worldviews in the context of everyday life', Nonreligion and Secularity Research Network blog, 4 October. Available at: https://nsrn.net/2016/10/04/methods-series-on-the-virtues-of-a-meaning-systems-framework-for-studying-nonreligious-and-religious-worldviews-in-the-context-of-everyday-life/ (accessed 21 February 2017).

Taylor, C. (2007) *A secular age*, Cambridge, MA: Harvard University Press.

Trzebiatowska, M. (2018) 'Sovereign of herself: women's narratives of "lived atheism"', *Secularism and Nonreligion*, 7(1): 1–11.

Trzebiatowska, M. and Bruce, S. (2012) *Why are women more religious than men?*, Oxford: Oxford University Press.

Zuckerman, P. (2008) *Society without God: What the least religious nations can tell us about contentment*, New York, NY: New York University Press.

9

# Avoiding the 'Anti-intellectual Abyss': How Secular Humanists in Sweden try to Define the Boundaries between Science, Religion, Pseudoscience and Postmodernism

*Susanne Kind*

## Introduction

In March 2015, the Swedish daily newspaper *Dagens Nyheter* published an article written by the former chairman of the Swedish Humanist Association (SHA) and other authors warning that the population of Sweden was being led towards an 'anti-intellectual abyss' (Ingvar et al, 2015b). This was one of several such statements co-authored by members of the SHA, who have recently stepped up their science-related activism to fight the phenomena they see as taking the country in this dangerous direction. This chapter analyses the SHA's activism and has three key objectives. First, it aims to provide insight into secular humanists' understanding of science. Second, it identifies the actors and phenomena that are perceived as 'others' of, and threats to, science by secular humanists. Finally, and most importantly, fulfilling the first two aims will shed light on the different ways in which secular humanists use references to science in their lobbying and activism, and draw attention to the internal tensions that are associated with these differences.

This study on which this chapter is based[1] is part of a growing research field of non-religion and secularity, which has gained increasing attention especially over the past 10–15 years. More concretely, it relates to research that emphasizes the added value of relational approaches[2] for exploring non-religious phenomena (Campbell, 1971; Quack, 2014; Wohlrab-Sahr and Kaden, 2014). While rooted in this field, the overlaps with science and technology studies will become apparent across the chapter (see Gülker, Chapter 5, this volume), especially studies of 'boundary work' to demarcate 'science' from 'non-science' (Gieryn, 1999; see also LeDrew, Chapter 10, this volume).

In addition to relational approaches to non-religion, this study has further been inspired by studies that focus on the importance of actors' constructions of collective identities, including the orientations of their actions in their respective social and cultural contexts (Melucci, 1996; Golova, 2011). Several studies point to the significance of science as one of several – and arguably the most central – reference points for diverse non-religious actors and an important element in their identity construction (Hess, 1993; Smith, 2011; Quack, 2012; Wohlrab-Sahr and Kaden, 2014). A few articles have even addressed SHA members' understanding of science, highlighting its central components (Lundborg, 2013). To date, however, no studies about humanism in Sweden have addressed the different ways in which secular humanists apply references to science, as well as the different relationships that secular humanists establish with counterparts in this context.[3] The present work thus complements existing research with its analysis of the different ways in which secular humanists construct their collective identity – something that has implications for their references to science and their relationships with other actors. The different ways in which secular humanists perceive their role, I will demonstrate, leads to diverging and sometimes competing forms of usage of science references within their campaigning.

This chapter is divided into five sections. First, I briefly elaborate on the theoretical assumptions that form the basis of the study. In the second section, I present the SHA as a collective actor and analyse members' self-image as an organization, introducing the ideal-typical concepts of 'life-stance humanism' (LSH) and 'opinion-making humanism' (OMH). This is followed by a third section in which SHA members' understanding of science (including the concrete content and forms of their references to science and the ascribed functions to science) will be delineated. In the fourth section, I illustrate and analyse for whom they use references to science: who are, from their point of view, the 'others' of science and, therefore, secular humanism? Finally, the concluding fifth section of the chapter deals with secular humanists' activism in the Swedish education sector in order to elucidate where they convey their scientific understanding and in which activities and projects their views of science manifest themselves. In this context, it will become clear that an emerging tension between LSH and OMH has a decisive impact on humanists' science-related activism.

## A relational approach to non-religious collective identities

I build on Quack's relational perspective – which is inspired by Bourdieu's methodological relationalism – as a fruitful avenue for the analysis of non-religious phenomena. He defines this as 'phenomena that are generally not considered religious but whose significance is more or less dependent on religion' (Quack, 2014: 439). He proposes a heuristic approach to further the understanding of relationships between 'religious' and 'non-religious' phenomena and actors. Importantly, 'religion' is not defined in advance: '[T]he "religious field" comprises all phenomena commonly understood to be religious in a specific socio-historical context, for example aspects of belief, behavior, and belonging attributed to religion or religions by the people being researched' (Quack, 2014: 448). Such an analysis involves looking at the positions of, for example, non-religious groups like the SHA, their specific historical, cultural and socio-political contexts, and the different ways in which they relate to certain religious others. This also includes taking into account the ways in which such actors evaluate and relate to the particular configurations of the respective religious field, including the legal regulations that, among other influences, shape such configurations. A relational focus differs from substantialist approaches to non-religion, which primarily highlight certain characteristics or elements of non-religious phenomena, and that tend to totalize the binary between religion and non-religion.[4]

Similar to Quack's approach to researching non-religion, Alberto Melucci also designed his approach to studying society relationally as a means to avoid essentialization. Melucci (1996: 67, 70) defines collective identity as 'an interactive process through which several individuals or groups define the meaning of their action and the field of opportunities and constraints for such an action', continuing that it is 'an interactive and shared definition produced by a number of individuals (or groups at a more complex level) concerning the orientations of their action and the field of opportunities and constraints in which such action is to take place'. In this sense, the following sections focus on the relational aspects of humanists' collective non-religious identity: in which ways do relationships to religious others influence humanists' construction of a collective identity, and their orientation towards, and positioning in, certain fields of action? What kinds of relationships with religious and other counterparts do humanists in Sweden aim for, which ones do they reject, and why?

A combination of Quack's relational approach to researching non-religion and Melucci's concept of collective identity facilitates an understanding of how actors focus on specific problems and goals, their meaning making, their actions, their network of diverse relationships, and, more generally, their orientation towards specific arenas and fields. Furthermore, it can provide important insight into how references to science are applied and framed, and the value that actors ascribe to them.

The following insights are based on empirical material consisting of qualitative data from 37 semi-structured interviews with members of the SHA, their opinion pieces in Swedish newspapers, field observations and publications in which SHA members introduce secular humanism, their goals as a humanist organization, associated key issues and their activities. Among the latter sources are the SHA's 'Programme of ideas' and a manual on secular humanism written by SHA members for teachers and students (Lindenfors and Sturmark, 2015, 2016). Both are representative of organized secular humanism in Sweden. The fieldwork included a pre-study in April and May 2013, and a 10-month period from August 2013 until May 2014 in the urban region of Stockholm and Uppsala. Interviews were coded and interpreted using the grounded theory approach (Corbin and Strauss, 1990), and with the help of the software MAXQDA.[5]

## The SHA: between life-stance organization and opinion-making interest organization

The SHA is the largest and most active association of 'secular humanists' in Sweden. Members of this association use 'secular humanism' as a self-designation to differentiate themselves from other forms of humanism, including religious forms, and those people in the Swedish context who are called 'humanists' for studying 'humaniora', for example, history, languages and philosophy (SHA, 2014a: 3).

The SHA was founded in 1979, in opposition to, in particular, the Church of Sweden and its position as state church at the time, but also to 'irrational' phenomena more generally. Since the beginning, the focus on religious criticism has been central to its members' activism. In the 1990s, members of the organization supplemented this focus by establishing humanist ceremonies, including non-religious ceremonies for newborns, confirmations, weddings and funerals, as an alternative to religious, especially Christian, offers. Today, the association is composed of 12 local organizations nationwide and has a total of about 5,300 members (Gunnarson, 2016; SHA, 2016: 1). Since March 2018, Anna Bergström, researcher in the field of global health, is chairwoman of the SHA. Before that, Christer Sturmark was chairman from 2005 until 2018. Sturmark is a well-known figure in Sweden for his work as a former information technology (IT) entrepreneur and his present work as a publisher,[6] author and regular participant in public debates. It is reasonable to assume that Sturmark's popularity and media presence is, in addition to the media presence of so-called New Atheism, one of the reasons for the fivefold increase in membership since the turn of the millennium.

One important aspect of the Swedish context is that, in connection with the separation of church and state in 2000, the Swedish government introduced legal regulations for the official recognition of faith communities (SFS, 1998). This status is a precondition for state support, in, for example, the form of financial aid. Funding is to be used to support faith communities to 'pursue active and long-term oriented religious activities in the form of worship, pastoral care [*själavård*], education and care' (SFS, 1999). Nilsson and Enkvist (2016) argue that this law, and the law for the Church of Sweden, illustrate the 'techniques of religion-making' and the 'performative role' of the Swedish state with regard to the recognition of faith communities. In the last section of the chapter, I point out that this 'performative role' also plays a crucial role for the identity construction of those secular humanists who aim for equal treatment with established faith communities.

On the basis of the analysis of the empirical material, I differentiate between two ideal-typical meanings of humanism within the SHA. First, 'life-stance humanism' (LSH) refers to defining secular humanism primarily as a comprehensive 'life stance' ('*livsåskådning*') and the SHA as a life-stance organization for secular humanists. Second, 'opinion-making humanism' (OMH) refers to defining humanism primarily as an opinion-making force and the SHA as an interest organization with a focus on religious criticism and human rights. The terms 'life stance' and 'opinion making', used by SHA members themselves, have been adopted to delineate two ideal-types of secular humanism, based on the analysis of the interviews with SHA members. The construction of these ideal-types helps to analyse and illustrate existing similarities and differences among secular humanists. The two ideal-types of humanism can be distinguished from each other in terms of the following:

1. activism, including (a) the objectives of collective activism, (b) the focus with regard to certain values and themes, and (c) the postulated scope of activism (eg individual lifestyles, society at large);
2. different types of relationships with religious actors;
3. organization, including (a) the pursued organizational form and (b) the self-image and perception of the organization, as well as its function; and
4. the orientation towards different social fields regarding the type of action, the institutionalization of secular humanism and the mobilization of resources.

In the organization's recent past, differences related to single aspects of these ideal-types led to heated debates among members. Understanding what these negotiations were about sheds light on what secular humanism in Sweden means to its representatives and, as will be shown in the last section of the chapter, helps us to understand the different roles that science plays for secular humanists' activism.

### *Secular humanism as a comprehensive life stance*

In the recent past, several SHA members have emphasized that humanism should be taken to the next level: towards a more comprehensive life-stance community, for example, strengthened through intra-organizational social activities and an expansion of humanist ceremonies (Gunnarson, 2016). A 'life stance' is defined 'as a basic understanding of the world that encompasses a descriptive (philosophical) and a normative (moral) dimension' (Sturmark and Börlin, 2012: 10). As such, the life-stance concept includes both 'religious' and 'secular' life stances, which share in common that they deal with ethical and existential questions about life, death, understandings of good and bad, and the beginning of the universe and our planet (Sturmark and Börlin, 2012).[7]

Advocates of LSH depict humanist ceremonial work as fulfilling the needs of people, who search for alternatives to traditional religious offers. Some that adhere to LSH emphasize that they should, as humanists, not allow faith communities to be 'the only game in town' (Damaschke and Ericsson Qvist, 2013: 17), which underscores the competitive, and in this sense relational, logic of this thinking. LSH advocates aim to establish an offer that is, in the sense of Stolz (2010), functionally as close as possible to religious offers (Gunnarson, 2009). Those in favour of LSH demand an expansion of the existing state regulations established for faith communities so that they are applicable to all life stances, religious and secular alike.

In summary, the ideal-type of LSH is characterized by:

1. the main objectives of (a) emancipation, empowerment and growth in order to strengthen the humanist movement and to fulfil the emotional and social needs of people through humanistic offerings, (b) an orientation towards, and dealing with issues and values that are related to, the relationship between secular and religious beliefs, in particular, their equality, and (c) the postulated scope for humanism covering society, the humanistic community/organization and individual everyday life;
2. functionally close competitive relationships with religious actors, with the aim of replacing them in their common-known functions;
3. representatives working to (a) establish humanism in the form of a life-stance community/organization, and (b) consider the community as having a value in itself; and
4. advocates orienting themselves towards the religious field, including legal regulations associated with it and the role, status and activities of faith communities.

## *Secular humanism as an opinion-making force*

Advocates of OMH want an organization categorically quite different from religious communities, one that focuses on opinion-making work and achieving common socio-political goals related to fighting religious oppression, defending secularism and human rights, and debating educational issues more generally. In 2012, internal debates arose around certain areas of activity and the SHA's general development. Some members expressed concerns about the SHA's ceremonial work, the understanding of humanism as a holistic life stance and the conception of the SHA as a life-stance community. They point out that a focus on the social collective and a shared life stance will have a negative impact on the emphasis on rational thinking and a genuine search for the best arguments. It would thereby undermine their norm-critical work and contradict what an interest organization should do. They wish for the SHA to stop organizing humanistic ceremonies as they do not consider the SHA to be an organization that should offer functional equivalents to religious services.

In summary, OMH as an ideal-type is characterized by:

1. the objectives of (a) defending and promoting secular values and the differentiation of politics and religion, (b) a primary focus on themes and values associated with universal, individual human rights, and (c) a postulated scope for humanism that extends to society and the humanist organization, but excludes individual lifestyles;
2. proponents engaging in relationships of criticism and conflict with religious actors, while rejecting functionally close competitions (functionally distant competitions and cooperative relationships are unproblematic as long as they do not presuppose a shared religion-like logic);
3. an organizational form that is (a) comparable to a political interest organization working for human rights and secularism, and (b) regarded as a means to achieve common goals; and
4. an orientation towards the wider political field, public discourse and alternative funding possibilities that does not relate to the status of faith communities, for example, for youth-related activities.

## SHA members' understanding of science

Despite such contestations about these aspects and their relation, science plays a key part in SHA members' overall portrayal of secular humanism. In their 'Programme of ideas', they dedicated a special section to 'Science, reason and critical thinking' (SHA, 2017: 4f), and in introductory lectures and study circles about secular humanism, science occupies a central position in relation to the 'descriptive' part of the secular humanist life stance, which refers to the 'naturalistic' perspective of how secular humanists perceive the world and reality (Sturmark, 2007: 48f).

Science is described as a method, an ongoing process, 'a structured and systematic testing and thinking' and 'a more systematic search for knowledge'[8] (Sturmark, 2015). In those documents in which SHA members explain what a secular humanistic life stance implies, they emphasize that secular humanists have a naturalistic view of the world and the universe. They hold the position that there are objective truths, for example, natural laws, that determine how the world physically works. Members of the SHA acknowledge that there are different views of the world, but in their view, only theories with good reasons to believe in them are worth taking seriously.

The scientific understanding and epistemological thinking of humanists have been influenced by various currents, among others, positivism, logical empiricism, Enlightenment ideas and naturalism.[9] This is shown by a combination of references to the ideas of, among other thinkers, Voltaire, David Hume, Immanuel Kant, Bertrand Russell and, importantly, also Ingemar Hedenius, a Swedish philosopher and a popular debater of the relationship between 'faith and knowledge'[10] (Lindenfors and Sturmark, 2015). In addition to these, references to the work of the researchers and public figures related to the American 'science wars' of the 1990s and so-called New Atheism are numerous. These include, among others, the work of Alan Sokal, Richard Dawkins, Christopher Hitchens and Daniel Dennett. In the context of discussing the idea that sensual experiences may deceive us, Lindenfors and Sturmark (2015: 77) point out that evidence occupies a central position. They link a high appreciation of empirical evidence with the significant value they place on reasonable and rational thinking.

Along with pointing to the limits of assessments that have been derived from sensory experiences, SHA members mention restrictions on the scope of science when it comes to aesthetic and moral judgements. They point out that science does not provide insights into 'how morality should be designed – how we should behave towards each other' – and 'how things should be' (Lindenfors and Sturmark, 2016: 24). Here, they draw an explicit boundary between science and morality. Under the heading 'The boundaries of science', they use humans' access to mathematical and philosophical knowledge through logical thinking as an example that '[t]here are other ways to gain knowledge than science' (Lindenfors and Sturmark, 2016: 24).

In contrast to these explanations about the boundaries of science, previous SHA descriptions of science hardly discuss the limitations of science, but instead appear to represent scientistic ideas: 'The experience-based reality is considered to be in principle explainable through scientific methods, although we will surely never understand it completely' (SHA, 2012a: 3). Here, science is positioned as the key to explaining all of reality, with the exception that humans might not be capable of reaching all explanations. However, this phrase has not been incorporated into more recent versions of the SHA's 'Programme of ideas', and annotations on the limitations of science have been included in the same and further publications (Lindenfors and Sturmark, 2016; SHA, 2017: 4). In sum, there seems to have been a shift away from descriptions of science that appear scientistic. In this regard, their understanding of science clearly differs from the positions of people who hold the view that science is the only way to understand reality (Dawkins, 2000) and acquire knowledge (Rosenberg, 2011).

There are no data about how many members are working as scientists, but there are already nine scholars from different Swedish universities among the 37 members whom I interviewed and the four persons who, together with the former SHA chairman, Sturmark, had published articles on postmodernist thinking. They research and teach in various natural and social science disciplines, among them physics, zoology, neuroscience, integrative medicine, philosophy, international health, organic chemistry and sociology. As such, they often take on the role of scientific experts in public debates and other contexts.

A declared goal of SHA members is to generate a larger public audience for science and current research through educational offers, such as public lectures by scientists from different disciplines. With the help of such and other public events – for example, visiting school classes – they present their understanding of science. One of their key demands in relation to science is that public decision-making and actions should, 'as far as possible, be based on knowledge and proven experience' (SHA, 2017: 5) and that knowledge, theories and practices that prove to be wrong and/or ineffective should not be publicly supported.

In interviews and publications, SHA members criticize, in particular, three categories of phenomena for questioning and contradicting a proven scientific understanding, for being irrational and for having negative social consequences: (1) religions and religious fundamentalisms that make truth claims; (2) pseudosciences; and (3) postmodernist and relativistic schools of thought.

## The 'enemies of the Enlightenment'

### *Religion and science: secular humanists as propagators of the conflict narrative*

SHA members emphasize that their naturalistic world view excludes the belief in a supernatural world and supernatural beings (SHA, 2014a: 3). While secular humanism and science are compatible, this does not, in their view, apply to science and religion, at least in cases where the latter includes truth claims. Numerous publications and YouTube videos show that SHA members actively participate in or even initiate public debates about the relationship between religion and science. One of their main arguments is that whereas science is based on a method and knowledge is subjected to experiments and observations, religions require trust in authorities, their description of supernatural existences and sometimes even the belief in things that 'our brains confirm to be purely wrong' (Lindenfors and Sturmark, 2016: 65). At some points, SHA members extend the scope of science beyond the possibilities of testability. They not only say that, according to knowledge-based standards and thinking, it is unlikely that, for example, miracles exists, but also impart the view that such beliefs conflict with existing scientific knowledge, that is, go against well-established knowledge (Lindenfors and Sturmark, 2012). Comparisons between scientific experiments and revelations, with regard to the question as to which offers the best source for knowledge, aim to show that religion cannot truly compete with science. In addition to emphasizing the antagonistic and conflictual relationship, such comparison highlights the competitive logic between science and religion, as perceived by secular humanists. Over the years, SHA members have criticized a variety of religious traditions and representatives of various faith communities and religious institutions, especially the Church of Sweden and its influence in Swedish society. Interestingly, there are also numerous contributions in which 'religion' is not further specified except for the reference to 'supernaturalism', a 'belief in God' or 'revelations'.

Religious beliefs are further perceived to pose a threat to the rational basis that public institutions and politics should be based on:

> Precisely because religious arguments are intellectually invalid, they should be kept outside of all public institutions' operations. The state and politically controlled organizations should be secular.... That religion – when used as an argument – is always a threat to rational discussion, of course, makes it necessary that it is questioned in many contexts. (Dannefjord, 2011: 65, author's translation)

In the eyes of Per Dannefjord, former SHA vice-chairman, something is irrational if it cannot be questioned and/or made the subject of experiments and proof. This applies not only to natural scientific facts, but also to certain forms of social communities, in particular, religious ones (Dannefjord, 2013). Similar to what Schlieter (2012) points out with respect to several authors who formulate naturalistic criticisms of religion (among them Richard Dawkins and Daniel Dennett), SHA representatives understand rationality only 'in a minimal sense (a) according to the scientific knowledge culture of verifying refutable statements ... as well as (b) in the sense of cognitive strategies for the solution of manageable tasks (i.e., instrumentalistic)' (Schlieter, 2012: 206, author's translation). SHA members' understanding of rationality has, to different degrees, been influenced by the British tradition of empiricism, as well as by ideas associated with rationalism. For SHA members, rationality is not oriented towards discourse, negotiation and communication, in which reasons can be delineated as plausible. Their perception of rationality can be distinguished from both a communicative rationality, as advocated by Habermas (1984), and from an understanding of rationality in the sense of a 'specific human form of reason (in the sense of classical metaphysics up to transcendental philosophy)' (Schlieter, 2012: 207, author's translation). It is universal rather than contextual.[11]

### *Science and its wannabes: secular humanists' efforts to limit the influence of 'pseudoscience'*

Next to religious truth claims and irrationality, 'pseudoscience' is one of the problematic trends with regard to science that SHA members invoke when they emphasize the need for the influence of humanism in Swedish society and beyond. They perceive these problematic trends as flourishing phenomena in today's world (SHA, 2017: 2). Thus, they refer to ideas and practices that delude people into thinking that their insights and practices are based on scientific methods and evidence, when they actually are not. They point out that a lack of empirical evidence alone is not the criterion for pseudoscience. All people believe in certain things without necessarily having the evidence to support the belief. For SHA members, the difference lies in claiming a scientific character while not meeting scientific criteria (Lindenfors and Sturmark, 2015: 79).

In order to show how prevalent pseudoscientific phenomena are in Sweden, SHA members refer to commercial contexts and business activities associated with 'New Age' practices, alternative medicine (eg homeopathy and halotherapy), astrology, spiritual healing practices, crystal energy, levitating Tibetan monks, intelligent design and creationism (Lindenfors and Sturmark, 2015; SHA, 2017: 4). The latter example clearly shows that there are no defined boundaries with regard to the concepts of 'religion' and 'pseudoscience', but, rather, that they are fuzzy and overlapping in the SHA's applications of them.

According to SHA members, pseudosciences are problematic because of their consequences. These are dangerous because they can cost people money and may even lead to serious mental and/or physical health damage. SHA members demand that social measures and political decisions – for example, medical and social services – should be based on scientific methods and proven experience. The marketing, selling and public funding of products that contradict scientific evidence and the effects of which cannot be proven scientifically ought to be prohibited (SHA, 2017: 5). In the opinion of SHA members, the reason for the growing influence of pseudoscience lies, among other things, in a (growing) lack of critical thinking, which, at the same time, is also the reason for why 'postmodernism' has spread within academia and among the wider population.

### *Postmodernism as a pivotal antagonist*

In recent years, SHA members have paid increasing attention to the themes they consider to relate to the current of 'postmodernism'. It is not a novelty that SHA members discuss 'postmodernism'; some references to it could already be found in earlier publications (Mozelius, 2003; Holst and Häggström, 2010). It is new, though, that postmodernist ideas, besides being regarded as highly irrational and socially problematic, are now classified as highly relevant to the Swedish public. What SHA representatives connect with this concept is essentially a deliberate choice or unconscious tendency of some people towards holding relativist positions. They reject different forms of relativism, including 'truth relativism', 'moral relativism', 'cultural relativism', 'fact relativism' and 'method relativism'.

The increased attention that SHA members pay to postmodernism manifests particularly in articles written in daily newspapers by members with headlines like 'The postmodern truth relativism leads us into an anti-intellectual abyss' (Ingvar et al, 2015a) and 'Pseudo-theories are equated with established science' (Danielsson et al, 2016). The latter article initiated a discussion on the debate pages of the daily newspaper *Dagens Nyheter* and became the subject of a public panel discussion on the question 'What is the situation with the view of knowledge in Swedish schools?'. SHA members object that there are pedagogues who put the theory of evolution on the same level as the Biblical creation story and treat them as 'two different paradigms', none of which is more 'true' than the other. In addition to pedagogues, the Swedish National Agency for Education (Skolverket), with respect to teaching material (Wikforss, 2016), and social scientists have been criticized for pleading for diverse forms of relativism in different contexts over the past 10–20 years, thus contributing to the dissemination of different relativisms. They fear that a point will come where claims dressed up as 'facts' compete with scientifically proven facts on a regular basis and where established scientific theories represent mere options among a 'smorgasbord' of theories (Danielsson et al, 2016). SHA members tend, then, to reduce 'postmodernism' to extreme forms of truth and value relativism.

These insights mainly illustrate abstract 'others' and phenomena that are regarded as problematic by humanists. SHA members also refer to specific people and examples when pointing to postmodernist tendencies in Swedish society, such as specific formulations in teaching material or the appointment of a new director of Södertörn University in Stockholm in 2010 (Holst and Häggström, 2010; Wikforss, 2016). However, there is a lack of strong counterpositions that enter into a continuing dialogue with the SHA regarding the assessment of these trends. In the past, counterparts openly debated SHA members and criticized their ideas and positions on various issues, including the Archbishop of the Church of Sweden, Antje Jackelén (2016) and the former chairman of the Swedish Evangelical Alliance, Stefan Gustavsson (2012). While SHA members refined their positions with respect to a few points of criticism, there is now, however, no strong opposition in Sweden, which, taken alone, has had a lasting impact on the positioning of SHA members. In the absence of a prominent religious counterpart, it is striking to what extent SHA members refer to situations and humanist organizations in other countries in order to legitimize the need for humanist activism more generally. Examples include the fight for liberal abortion laws in Poland, against the caste system in India and against witchcraft in African countries (Sturmark, 2016). By contrast, recurring, central points of reference for humanist criticism and goals in Sweden are the laws on the official recognition and public funding of faith communities, mentioned earlier. These regulations have had a lasting effect on the positions of humanists, both adherents to LSH and OMH, and will therefore be discussed with respect to their influence in the following sections.

## Non-religious collective identities and their impact on science references

SHA members are 'propagators of a rational-scientific world-view' (Campbell, 1971: 7), but, importantly, they act as such in quite different ways. Over the course of my research, it appeared that while SHA members seem to largely agree in their basic understanding of science, the differences between LSH and OMH decisively influence how SHA members apply and frame references to science. Among the key factors are the different kinds of relationships that humanists engage in and strive for with state-recognized religious communities, and their relationship to the representation of 'world religions' in different contexts, such as school textbooks. This became clear with regard to different activities in the education sector, as the following examples epitomize.

Advocates of LSH have several goals in the field of education. Among other things, they demand a comprehensive and accurate inclusion of the secular humanist life stance in religious education ('*religionskunskap*'). In addition, they propose that religious education be renamed and re-conceptualized into 'life-stance studies' ('*livåskådningskunskap*'), a subject that deals with the great questions of life (Sturmark and Börlin, 2012). The primary goal is that secular humanism attains an equal position in relation to religious beliefs. In this context, secular humanists compete with religious actors, who have an established position in textbooks and the syllabus. For the inclusion and presentation of secular humanism, proponents of LSH focus on religious education and its transformation, a framework that provides a logic of functionally similar and partly equivalent life stances.

The teachers' magazine *Religion and Questions of Life* (*Religion och Livsfrågor*), issued by the Association for Teachers of Religious Education (Föreningen Lärare i Religionskunskap), devoted a whole issue to 'secular humanism' in 2012, in which Sturmark, the SHA's former chairman, and SHA member Anita Börlin describe secular humanism and the SHA in the following way:

> In Sweden, people with a secular humanistic life stance organize themselves in the association Humanisterna [SHA], just as a certain type of Christian faith become members of the Pentecostal Church, or people with Muslim faith are members of a Muslim community. (Sturmark and Börlin, 2012: 10, author's translation)[12]

In this context, references to science are primarily used as an identifying feature of the secular humanist life stance and a distinctive characteristic, marking one of the differences between 'religious' and 'secular' life stances and associated organizations. Those in favour of LSH place references to science in functionally close competition with 'religious' life stances. A teacher of religious education told me that he explicitly invites SHA members to represent a secular life stance that broadly relies on and refers to science. He planned to include podium discussions in his courses in which secular humanists, as representatives of the Darwinian theory of evolution, debate with creationists about their respective positions and beliefs.

A Swedish textbook also describes secular humanism as an example of those life stances that are 'built around the scientific knowledge we have about reality' (Björlin and Jämterud, 2013: 333). The authors of this textbook, who have been advised by Sturmark, further elaborate that secular life stances differ from religious ones in that they are based on references to science while the basis for religious beliefs is their respective faith:

> Secular life stances thus have a view of the world and humans based solely on what we can comprehend through reason and what science seems to know, emphasizing that it is this, along with a secular foundation for ethics, which is the sole foundation for a sustainable view of life. However, this does not mean that you have a 'religious' worship of reason, but only that you think it's impossible to build your life on things that do not meet reason's truth claims. (Björlin and Jämterud, 2013: 342f, author's translation)

Furthermore, they rather implicitly point to the fact that references to science, as a part of a life stance, acquire some kind of a world view character themselves: 'Secular humanists see their life stance as a belief in science and rationalism (belief in reason)' (Björlin and Jämterud, 2013: 346).

Although these activities are carried out in the field of education, here, proponents of LSH appear as the representatives of a secular life stance and as an alternative to religious ones. Thus, they are clearly portrayed in relation to religious actors, through competitive relationships and to the religious field more generally. Field-theoretically, they play a shared 'game' with religious life stances and envision the religious field to be extended and transformed into a life-stance field. Members explicitly emphasized the strategic possibilities that presentations of secular humanism in the school sector offer, particularly when it comes to increasing awareness of humanist ceremonies. Scientific references serve as identity markers and are used to promote secular humanism as a rationalist life stance that grasps reality in a better way than outdated revelations.

In addition to this representation in textbooks, secular humanists in Sweden have also been successful in moving towards equal recognition of their organization in relation to religious communities such as the Church of Sweden. Since 2012, for example, there has been a secular inauguration ceremony for MPs as an alternative to the traditional opening ceremony organized by the Church of Sweden. For more than

10 years, the SHA has regularly applied for state recognition on the same legal basis as faith communities. So far, all applications have been rejected on the grounds that the SHA does not have a transcendental dimension and does not organize religious activities, but rather distances itself from such activities (Kulturutskottets betänkande, 2006/07: 9f). The SHA members, in particular, supporters of LSH, argue, in turn, that they perform functions for their members and society similar to the ones of faith communities. Among other things, they attempt to illustrate this similarity through linguistic rapprochement, pointing out that their activities also include counselling activities ('*själavård*') similar to pastoral care in faith communities (Sturmark et al, 2012). The Committee on Cultural Affairs' (Kulturutskottet's) rejection of SHA applications and the argumentation of SHA representatives show the performative role that the Swedish state plays with respect to the making of religion and non-religion (Nilsson and Enkvist, 2016).

In contrast to such an LSH focus and associated role of humanism as a functional equivalent of religion, advocates of OMH within the SHA primarily emphasize addressing problems and questions about social values as a motivation for their activism. In view of the challenge associated with defending critical thinking against unscientific, irrational and immoral approaches towards knowledge and universal values, members of the SHA partly extended their activism. The conception and implementation of their projects in the field of education are relatively independent of the subject of religious education and the associated concept of comprehensive life stances. They object, for example, that critical thinking is not taught comprehensively enough in Swedish schools and that it is not clearly defined in curricula. According to them, this is one of the reasons why irrational postmodernist tendencies are gaining ground in the educational field.

Their activities include training courses for teachers, one of which explicitly deals with critical thinking. They are working on a manual to provide teachers and students with ideas on how to teach critical thinking in all school subjects, which members of the SHA internal school group, founded in 2012, told me about in interviews. Another example is the project 'Can we talk about this here?' ('Kan vi prata om det här?'), which was carried out at several Swedish schools in 2012. It aims, among other things, to discuss the negative consequences of cultural relativism, which is described as a very influential perspective 'that controls how some think and argue about human rights' (SHA, 2012b: 5). According to the authors, such a cultural-relativist perspective means:

> that only those belonging to the group can criticize it and its actions, all external criticism is thus unwelcome. The opposite of cultural relativism is universalism. A universalist point of departure means that human rights apply to all people irrespective of culture, ethnicity and geographical residence. (SHA, 2012b: 38)

SHA members argue that debates about themes relating to racism and intolerance show that there exist 'clearly defined knowledge gaps' when it comes to the themes 'cultural relativism, boundary drawing, group versus individual perspectives, and tolerance versus intolerance' (SHA, 2012b: 37). These topics are perceived as being absent in teaching as well as public debate (SHA, 2012b: 36f). This project is an example of where SHA members see voids that they want to fill and where opinion-making work and the promotion of critical thinking are deemed necessary. The focus on discussing social questions and human rights issues more generally with students, independent of a narrow life-stance frame, also opened up new possibilities for financing. OMH supporters reject applying for state funding on the basis of equal treatment with recognized faith communities, but this project has been funded by the Swedish Agency for Youth and Civil Society (Myndigheten för ungdoms- och civilsamhällesfrågor), a government agency.

It is striking that these projects focus not on the equal treatment of and comparison between secular and religious life stances, but on the conveyance of values through critical examination of the relationship between cultural-relativistic attitudes and universal human rights and of a 'scientific approach' towards value and factual questions in general. Proponents of OMH therefore act as advocates and guardians of scientific theories and institutions, as well as scientific and critical thinking and their intrinsic and social value in a more general sense. They aim to strengthen, promote and defend scientific and critical thought against a variety of threats and the actors who represent, in the OMH view, such risks and negative influences. As a result, all three of the earlier-mentioned categories of phenomena and counterparts classified as irrational come into view with regard to their activism, rather than primarily focusing on established religious life stances as the main 'irrational' antagonist.

These projects establish relationships of criticism, conflict and functionally distant competition – for example, about values – with actors presenting religious, pseudoscientific and postmodernist ideas. In contrast to the activism of LSH advocates, it is not the aim of

OMH supporters – they explicitly reject it – to establish themselves as functionally close competitors with an end goal of substitution. In their recent activism, they primarily addressed representatives of the National Agency for Education (Skolverket) and pedagogues at school and university level who, in their view, represent postmodernist thinking or at least consciously or unconsciously grant such thinking space in the educational sector.

Here, 'science' does not serve as an identifying feature of a 'science-based life stance', but scientific references are used to defend evidence and rationality as the means to prevent the spread of anti-scientific tendencies. Instead of emphasizing the role of science as the basis for the secular humanistic life stance, OMH stresses the value of science itself: 'the best tools for understanding reality and seeking answers to questions are provided by science and rational thinking' (SHA, 2014b: 6). In this context, SHA members act primarily as members of an interest organization with political interests and, at times, as scientific experts. In contrast to supporters of LSH, they do not appear as advocates of a secular life stance as an alternative to religious ones.

While these examples show the different activities of SHA members in the education sector, the following critique illustrates why some members perceive these activities and underlying understandings of humanism as conflicting and mutually exclusive. Sören Holst (2013), a scientist in the field of physics, criticizes the LSH approach for 'largely mimick[ing] the way in which religion play[s] our emotions, with everything from wishful thinking to rituals [*hyllningsritualer*]'. He identifies the development towards a life stance with the creation of a community, a sense of belonging and the organization of ceremonies. While all people should have the right to use ceremonies, as they have 'the right to read horoscope[s]', he considers it a 'mistake' to incorporate statements like 'I believe in humans' into humanism and to develop ceremonies for celebrating 'the four turning points of life'. He juxtaposes the focus on criticism of religion and the work for a secular society with the search for togetherness in a life-stance community, and describes the latter compared to the former as 'something completely different'. He believes that most SHA members have joined the organization to support criticism of religion and to work for a secular society, not to find togetherness in a shared life stance. In his view, the latter leads to 'life-stance relativism' instead of universality.

He names several examples of such a life-stance framing, including applying for the same state funding as faith communities, being included into textbooks for religious education 'side by side with religions', arguing and applying for the entitlement of 'humanist celebrants' to wed people, and articulating a rhetoric that explicitly states that they have everything that religions have 'but without God'. He summarizes that:

> [a]ll in all, this leads us to emerge as an alternative religion, which is not only conspicuous in itself, it is a misconception of the same kind and with the same serious consequences as when the evolutionary teaching in the United States is presented as 'one theory among others'. (Holst, 2013)

Other proponents of OMH expressed their perception of internal conflicts with respect to a life-stance framing in the interviews with phrases like "the issue splits the organization in half" and those are "fields that shouldn't be mixed". One of my interviewees left the organization because of the life-stance framing and associated positioning as a functional equivalent to religion. Other members told me about further former members who reportedly left the SHA for the same reason.

As Quack (2014: 452) points out, relational studies of non-religion are particularly well suited to study the 'elective affinity' between non-religious actors' rejection of different phenomena. The example outlined previously illustrates that advocates of OMH criticize the positioning and framing as one of a variety of life stances for partly the same reasons that they criticize postmodernism, namely, that both render universal values and scientific criteria an alternative among numerous options.

## Conclusion

Advocates of humanism as a life stance and humanism as an opinion-making force agree on the use of science as an argument against religion, but they do not agree on the question of whether they as an organization should communicate science as a constitutive element of secular humanism's comprehensive secular life stance – a collectively organized *Weltanschauung* in which science serves 'as a means of giving meaning to life' (Kaden and Schmidt-Lux, 2016: 73). SHA members act as promoters and guardians of science, scientific methods and universal humanistic values, but despite the seemingly shared understanding of science, their methods and aims differ.

LSH and OMH science-related activism are guided by diverging constructions of non-religious identities, including orientations towards different social fields for their respective collective action.

As this chapter describes, this also implies that representatives of both ideal-types engage in different kinds of relationships with differently constructed counterparts. LSH partially are, and seek to further become, part of a field that has been established for and by faith communities and to transform it into a 'life-stance field'. They present secular humanism as functionally equivalent, but distinct in content, to religious life stances. In this context, which is shaped by the comparison of and the competitive relationship between secular and religious life stances and the logic of their fundamental conceptual similarity, the role of science for the secular humanistic life stance is emphasized and science is used as an identifying and distinguishing feature vis-a-vis competitors. Representatives of OMH fundamentally reject framing humanism in this way. They reject such framing, including its perceived consequence of 'life-stance relativism', and postmodernism for partly similar reasons – both potentially lead to the point at which scientific standards and arguments, as well as universal values, are interpreted as one of many options, undermining their universal validity and legitimacy. In addition, positioning as a life-stance organization among many would mean that their arguments concerning scientific topics, as well as human rights issues, and their role as an interest organization in this field, would no longer be taken seriously and would lose their credibility. According to them, there exists a logical tension between the self-image, goals, activities and, thus, positioning of SHA members who understand and position secular humanism as a life stance in Swedish society and their concerns as secular humanists.

Those who subscribe to OMH see themselves as an interest group, an opinion-making force, as part of a broadly conceptualized political field in which they act as self-proclaimed guardians of science and struggle with other actors about the legitimate assertion of the right understanding of science, critical thinking and rationality as the basis for other societal fields, like the educational one. From a field-theoretical perspective, OMH proponents argue that their activism as humanists should follow a logic that is fundamentally different from the logic that governs the workings of actors in the religious field. They refuse to enter into functionally close competition with religious communities, which presupposes that they play the same 'game' and follow the same rules.

While advocates of LSH use the field of education primarily as a battleground for their competition with religious life stances, those of OMH use it primarily more broadly as an arena for battles about secularism, critical thinking and views of knowledge, science and values. The latter especially focus on perceived problems in relation to postmodernism and cultural relativism, as well as a lack of critical thinking, whereas the former fight the unequal treatment of religious and secular life stances. For the former, then, science is a tool used in a game with religious competitors, while for the latter, a claim upon science is central to a sustained struggle to change the rules of the game itself.

## Notes

1 This chapter is based on my doctoral dissertation project on non-religion in Sweden. Parts of the chapter were written within the context of the Emmy Noether Project, 'The Diversity of Non-religion' (QU 338/ 1-1), funded by the German Research Foundation (DFG).

2 To take a relational approach means to analyse the positions of actors and institutions in social space, thus to take into account the dynamics and power relations between social positions (Bourdieu and Wacquant, 1992).

3 There are, however, studies that use relational perspectives to examine humanism in other national contexts, such as Kaden (2018), who illustrated the relationship between American humanism and the sociology of religion, and Schröder (2013), who analysed the involvement of humanist organizations in inter-religious initiatives in Germany.

4 Examples of substantialist understandings of non-religion include Zuckerman's (2008: 96) study of godless Scandinavian societies and Lee (2015: 35), who distinguishes between religious and non-religious 'cultures' such as humanism and political secularism based on the presence of 'strong non-religious components' (see also Lee, Chapter 8, this volume).

5 Interview questions focused on humanists' self-understanding, their relationship to conflictual issues within the organization, the role and function of their organization in Swedish society, and their relationships with other actors. An inductive approach, with open questions, was followed by a series of alternating inductive and deductive steps, which served to generate hypotheses and to guide data collection. The codes were developed on the basis of the empirical material. After a primary open coding, the subcategories mentioned at the end of the following section were analysed and coded, in the sense of axial coding, and afterwards scrutinized with respect to their relations to each other and the two key categories LSH and OMH, in the sense of selective coding (Corbin and Strauss, 1990).

6 Sturmark is head and chief editor of the publishing house Fri Tanke. Among other things, they publish Swedish translations of books by 'New Atheists', such as Richard Dawkins, Christopher Hitchens, Sam Harris and Lawrence Krauss.

7 SHA members use the term 'life stance' synonymously with the term 'world view', with life stance (*livsåskådning*) being the more commonly used expression.

8 All quotations from SHA members' publications and public statements have been translated into English by the author.

9 Debates on their blog, *humanistbloggen*, clearly show that the exact meaning and associated content of those currents are matters of controversy.

10 With his book of the same name, he triggered one of the largest cultural debates in mid-20th-century Sweden (Hedenius, 1949).

11 However, there are single references from other SHA members that suggest a different understanding of rationality. Thus, Gunnarson (2012: 12) points out that the satisfaction of people's emotional side can also be described as rational.

12 In contrast to those SHA members who focus primarily on LSH or OMH, the SHA's former chairman positions himself in a way that allows supporters of both sides to identify with certain of his positions and activities while critically judging others.

## References

Björlin, O. and Jämterud, U. (2013) *Under samma himmel. Religionskunskap för gymnasieskolan*, Stockholm: Sanoma Utbildning.

Bourdieu, P. and Wacquant, L.J.D. (1992) *An invitation to reflexive sociology*, Chicago, IL: University of Chicago Press.

Campbell, C.D. (1971) *Toward a sociology of irreligion*, London: Macmillan.

Corbin, J. and Strauss, A. (1990) 'Grounded theory research: procedures, canons and evaluative criteria', *Zeitschrift für Soziologie*, 19(6): 418–27.

Damaschke, R. and Ericsson Qvist, J. (2013) 'Till gemenskapens försvar', *HumanistInfo. Humanisternas medlemstidning*, 15(1): 16–17.

Danielsson, U., Moberg, C., Sturmark, C. and Wikforss, Å. (2016) 'Pseudoteorier jämställs med etablerad vetenskap', *Dagens Nyheter*, 11 January. Available at: www.dn.se/debatt/pseudoteorier-jamstalls-med-etablerad-vetenskap/ (accessed 2 October 2018).

Dannefjord, P. (2011) 'Gerles intellektuella haveri', *SANS. Existentiellt Magasin I Upplysningens Anda*, 1(1): 64–7.

Dannefjord, P. (2013) 'Gemenskapens Baksidan', lecture in Kalmar, 17 February. Available at: www.youtube.com/watch?v=V7NmE80QS8E (accessed 2 October 2018).

Dawkins, R. (2000) 'Thoughts for the millennium', Microsoft Encarta Encyclopaedia.

Gieryn, T.F. (1999) *Cultural boundaries of science: Credibility on the line* (2nd edn), Chicago, IL: University of Chicago Press.

Golova, T. (2011) *Räume kollektiver identität: Raumproduktion in der »linken szene« in Berlin*, Bielefeld: Transcript Verlag.

Gunnarson, S. (2009) 'Ceremonierna ger mänskligt ansikte åt humanism', *Humanisten. Tidskrift för Kultur- och Livsåskådningsdebatt*, 15(4): 26–9.

Gunnarson, S. (2012) 'Historiska perspektiv på humanismen: Del 4. upplysning, kritik och nydaning', *HumanistInfo*, 14(4): 12–13.

Gunnarson, S. (2016) 'Humanisterna har nu chansen att ta nästa steg', *Humanisten. Humanisternas Medlemstidning*, 17(2): 11–12.

Gustavsson, S. (2012) 'Humanisterna och synen på vetenskap', Blog Samtid och Sanning. Available at: www.stefangustavsson.se/?p=1372 (accessed 2 October 2018).

Habermas, J. (1984) *The theory of communicative action, vol. 1: Reason and the rationalization of society*, Boston, MA: Beacon Press.

Hedenius, I. (1949) *Tro och vetande*, Stockholm: Bonnier.

Hess, D.J. (1993) *Science in the new age: The paranormal, its defenders and debunkers, and American culture*, Madison, WI: University of Wisconsin Press.

Holst, S. (2013) 'Humanismen – en tro bland andra?', *HumanistInfo*, 15(2): 12.

Holst, S. and Häggström, O. (2010) 'Varför vill Södertörns högskola ha en vetenskapsfientlig rektor?', *Newsmill*, 7 January.

Ingvar, M., Sturmark, C. and Wikforss, Å. (2015a) 'Den postmoderna sanningsrelativismen leder oss ner i en antiintellektuell avgrund', *Dagens Nyheter*, 22 March.

Ingvar, M., Sturmark, C. and Wikforss, Å. (2015b) 'Halvförstådda idéer påverkar offentliga debatten', *Dagens Nyheter*, 26 March.

Jackelén, A. (2016) *Gud är större*, Lund: Arcus Förlag.

Kaden, T. (2018) 'American humanism and sociology of religion', in P. Ulrich Merz-Benz and P. Gostmann (eds) *Humanismus und Soziologie*, Wiesbaden: Springer VS, pp 161–81.

Kaden, T. and Schmidt-Lux, T. (2016) 'Scientism and atheism then and now: the role of science in the Monist and New Atheist writings', *Culture and Religion*, 17(1): 73–91.

Kulturutskottets betänkande (2006/07) *Stöd till ungdomsorganisationer och trossamfund*, KrU11. Available at: https://data.riksdagen.se/fil/21CC9318-CE00-4F52-986B-EECE28AF2008 (accessed 22 January 2019).

Lee, L. (2015) *Recognizing the nonreligious: Reimagining the secular*, Oxford: Oxford University Press.

Lindenfors, P. and Sturmark, C. (2012) 'En myt att vetenskap kan förenas med religion', *Svenska Dagbladet*, 11 December.

Lindenfors, P. and Sturmark, C. (2015) 'Humanism som livsåskådning. Förnuft, omtanke, ansvar. Studiematerial för grund- och gymnasieskolan från förbundet Humanisterna', long version.

Lindenfors, P. and Sturmark, C. (2016) 'Humanism som livsåskådning. Förnuft, omtanke, ansvar. Studiematerial för grund- och gymnasieskolan från förbundet Humanisterna' short version. Available at: http://media.humanisterna.se/pdf/humskola.pdf

Lundborg, J. (2013) 'Från hedenius till humanisterna', in C.R. Bråkenhielm, M. Essunger and K. Westerlund (eds) *Livet enligt människan: Om livsåskådningsforskning*, Nora: Nya Doxa, pp 273–92.

Melucci, A. (1996) *Challenging codes. Collective action in the information age*, Cambridge: Cambridge University Press.

Mozelius, P. (2003) 'Postmodernism – sekulär mysticism', *Humanisten: Tidskrift för Kultur- och Livsåskådningsdebatt*, 9(4): 18–22.

Nilsson, P.-E. and Enkvist, V. (2016) 'Techniques of religion-making in Sweden: the case of the Missionary Church of Kopimism', *Critical Research on Religion*, 4(2): 141–55.

Quack, J. (2012) *Disenchanting India: Organized rationalism and criticism of religion in India*, New York, NY: Oxford University Press.

Quack, J. (2014) 'Outline of a relational approach to "nonreligion"', *Method and Theory in the Study of Religion*, 26(4/5): 439–69.

Rosenberg, A. (2011) *The atheist guide to reality: Enjoying life without illusions*, New York, NY: W.W. Norton Company.

Schlieter, J. (2012) 'Der rationalitätsbegriff naturalistischer religionskritik', in U. Berner and J. Quack (eds) *Religion und kritik in der moderne*, Berlin: Lit Verlag, pp 197–220.

Schröder, S. (2013) 'Dialog der weltanschauungen? – Der humanistische verband deutschlands als akteur im interreligiösen dialoggeschehen', in S. Führding and P. Antes (eds) *Säkularität in religionswissenschaftlicher perspektive*, Göttingen: V. and R. Unipress, pp 169–85.

SFS (Svensk författningssamling) (1998) 'Lag om trossamfund', Svensk författningssamling, 1998:1593.

SFS (1999) 'Lag om stöd till trossamfund', Svensk författningssamling, 1999:932.

SHA (Swedish Humanist Association) (2012a) 'Programme of ideas', Stockholm: Swedish Humanist Association.

SHA (2012b) *Kan vi prata om det här?* ['Can we talk about this here?'], Stockholm: Swedish Humanist Association. Available at: http://media.humanisterna.se/pdf/skolmanual.pdf (accessed 22 January 2019)

SHA (2014a) 'Studiecirkel material'. Version from 2015 available at: http://humanisterna.se/wp-content/uploads/2017/03/Kompendium_2015.pdf (accessed 22 January 2019)

SHA (2014b) 'Att göra skolbesök', Stödmaterial. Stockholm: Swedish Humanist Association.

SHA (2016) 'Kongresshandlingar'. Stockholm: Swedish Humanist Association.

SHA (2017) 'Programme of ideas'. Stockholm: Swedish Humanist Association. Most recent version is available at: http://media.humanisterna.se/pdf/ip.pdf (accessed 22 January 2019).

Smith, J.M. (2011) 'Becoming an atheist in America: constructing identity and meaning from the rejection of theism', *Sociology of Religion*, 72(2): 215–37.

Stolz, J. (2010) 'A silent battle: churches and their secular competitors', *Review of Religious Research*, 51(3): 253–76.

Sturmark, C. (2007) *Tro och vetande 2.0: Om förnuft, humanism och varför människor tror på konstiga saker*, Nora: Nya Doxa.

Sturmark, C. (2015) *Upplysning i det 21:a århundradet. Tro och vetande 3.0*, Stockholm: Fri Tanke.

Sturmark, C. (2016) 'Är religion eller sekulär humanism bäst för vårt samhälle?', panel debate, ABF-huset Stockholm, 17 March. Available at: www.youtube.com/watch?v=y73PgfBVFKE (accessed 2 October 2018).

Sturmark, C. and Börlin, A. (2012) 'Sekulär humanism – i skolan och livet', *Religion and Livsfrågor*, 41(4): 10–11.

Sturmark, C., Westerberg, B., Fragell, L. and Liljevall, D. (2012) 'Se över lagen om statligt stöd till trossamfund', *Dagens Nyheter*, 13 June.

Wikforss, Å. (2016) 'Hur står det till med synen på kunskap i den svenska skolan?', 13 April. Available at: www.youtube.com/watch?v=5dBYFVkSxmQ (accessed 2 October 2018).

Wohlrab-Sahr, M. and Kaden, T. (2014) 'Exploring the non-religious: societal norms, attitudes and identities, arenas of conflict', *Archives de Sciences Sociales des Religions*, 167(3): 105–25.

Zuckerman, P. (2008) *Society without God: What the least religious nations can tell us about contentment*, New York, NY: New York University Press.

10

# Atheism and the Social Sciences

*Stephen LeDrew*

## Introduction

In the past decade, the anti-religious movement known as the New Atheism has been working to define a scientific basis for opposing religion and its influence in public affairs. Their thought on religion is rooted in scientism and a narrative of progress and Enlightenment that refers to ideas derived from evolutionary biology for its authority. A notable feature of this approach is the New Atheism's critique of the social sciences, which it dismisses as a relativistic domain of knowledge associated with postmodernism, and seeks to replace with evolutionary psychology and neuroscience. This chapter examines the New Atheism's position in terms of a historical division within atheist thought between approaches to religion that are grounded in the natural and social sciences, and argues that contemporary atheism should be understood primarily as a political project to advance the authority of an ideological vision of 'true' science and its representative experts.

The New Atheism is a cultural and intellectual movement originally consisting of the 'Four Horsemen' (as they were dubbed by their admirers): evolutionary biologist Richard Dawkins; philosopher Daniel Dennett; journalist Christopher Hitchens; and Sam Harris, a public intellectual with no significant credentials before he wrote *The end of faith* in 2004 (Harris, 2004), which became an unexpected bestseller and kick-started the movement. A number of other figures came into the fold, mostly consisting of professional scientists and public intellectuals, and a new social movement coalesced around their ideas that sought a place for atheists in public discourse (LeDrew, 2015).

While typically understood as a response to religious fundamentalism, the New Atheism is as much a reaction to a perceived weakening of universalistic standards of knowledge and morality in an increasingly pluralistic Western cultural milieu. Most of the criticism of the New Atheists has concerned their ideas about religion, including some gross generalizations about believers, and a generally hysterical attitude towards Islam and practising Muslims in particular, which are presented as a dire existential threat to Western civilization. The New Atheists' ideas about the social sciences have not received as much attention, though their books and public comments contain a strong and consistent critique of sociology and philosophy, in particular, as bastions of relativism that undermine scientific authority and, by extension, the entire project of liberalism, which they see as a direct outcome of scientific progress. In what follows, then, I show how New Atheism emerged out of a tradition of 'scientific' atheism that contrasts with an 'anthropological' or 'humanistic' atheism, before explaining how this contrast shapes the New Atheists' critique of the social sciences and their efforts to demarcate the boundaries of legitimate science.

## Modern atheism and the sciences

The history of atheism is complex and highly contested. Depending on one's point of view, atheism could be: the product of the turn to reason and empiricism instigated by the Enlightenment; a radical political challenge to the oppressive forces of church, state and monarchy; or the natural outcome of the evolutionary trajectory of culture and belief from supernaturalism to a scientific world view. Atheism is certainly not a new phenomenon, as highlighted by Tim Whitmarsh's (2015) recent study of atheism in ancient Greece. (In fact, Whitmarsh argues that atheism is as old as Abrahamic monotheism itself.) However, in the modern period, atheism took on a distinct form that reflected the evolving intellectual and cultural environment. Michael J. Buckley's (1987, 2004) books on atheism in the modern world are still definitive treatments of the subject that are shockingly overlooked in the field. Buckley outlines how a distinctly modern form of atheism developed immanently from within theology as a result of its response to Enlightenment criticisms, and how another form of atheism concerned with anthropological questions arose at roughly the same time that differed from the science-based approach of

Enlightenment critics. Understanding this historical distinction is key to understanding the present situation, where organized atheism today embraces one of these traditions, while the 'anthropological' approach is almost entirely neglected, reflecting the general antipathy towards the social sciences expressed by today's most popular exponents of atheism.

Buckley argues that a revolution within Christian theology coincided with the Scientific Revolution. Seeking an empirically grounded defence of the existence of God, theologians embraced science – contrary to popular belief – as a means of verifying God's presence in nature. The conception of God thus transformed from one of an immaterial, transcendent spirit, to a material presence in the universe. Gavin Hyman (2007) notably supports Buckley's interpretation, arguing that in the 17th century, a new, modern conception of God emerged that viewed God as a thing of physical substance and definite location within the universe. As a consequence of this view, God became an object of scientific inquiry. The unintended effect of this move was that rather than confirming the existence of God, science seemed more and more to indicate that God was, at best, an unnecessary addition to scientific theories of nature, and, for many, came to be seen as a direct contradiction of emerging scientific knowledge. Buckley therefore concludes that theology was ultimately responsible for the rise of atheism by creating a conception of God that could be challenged by critiques based in empiricism and logic, namely, those that emerged from the Enlightenment.

Despite this development, atheism as we know it today did not really come into its own until the mid-19th century due to the lack of a scientific theory that could account for the origin of life. As long as this was missing, there was a role for God in nature, and science-based critiques were limited. This changed with Darwin's theory of evolution by natural selection, the signature event and idea in the history of modern atheism. With the discovery of evolution by natural selection, atheists were armed with what they considered a complete set of theories and facts that explained nature with no need to invoke a deity. The argument of design lost considerable power as evolution gained traction in the intellectual world and society more generally.

Evolution was the trigger that launched a fully developed critique of religion, and a systematic belief system in its own right, which I call scientific atheism (LeDrew, 2012). A fusion of Enlightenment rationalism and Darwinism, scientific atheism understands religion to be a false explanation of nature that has been replaced by modern science. A teleological view of history is at work in this idea of intellectual progress and accompanying vision of social progress. Religion is positioned as a relic of ancient superstition and barbarism, while science lights the path to civilization, and is assumed to drive social and moral improvement. Crucially, it is the natural sciences that occupy the privileged position in this belief system, being the presumed antithesis of religion and the inheritor of the sole right to explain nature (science's proper province), but also to explain the development of human civilization. Scientific atheism was intimately wedded to Darwinism and, by extension, to the effort to emancipate science from the conservative force of the church (Olson, 2008). It was thus expressed most forcefully, perhaps, by Thomas Huxley – who actually declared himself an agnostic, though his public criticisms of theology and orthodoxy were as forceful as the New Atheists' criticisms of religion are today (see Lightman, 2001) – who took on the role of public defender of evolution against its religious critics. Huxley was a fierce advocate of liberalism, which, for him, went hand in hand with a scientific, non-religious world view. From its beginnings, then, scientific atheism was as much political ideology as epistemology.

While the scientific tradition came to dominate atheist thought and activism, there is another major branch of atheism that has been very influential at times in the past. David Berman (1988) refers to the 'anthropological approach' to atheism, which, in his view, was pioneered by Ludwig Feuerbach and Karl Marx, and later included thinkers such as Friedrich Nietzsche and Sigmund Freud. I call this 'humanistic atheism' (LeDrew, 2012) in reference to its emphasis on the fact that religion responds to human needs that are both social and psychological in nature. The chief distinction between this and the scientific approach, then, is that humanistic atheism recognizes that religion is a social phenomenon, and not solely a pseudoscientific explanation of nature. It is grounded in the social sciences and philosophy, rather than the natural sciences. While it comprises a diverse group of thinkers, this approach to religious criticism is characterized by the recognition that religion is both a consoling and an oppressive force.

This point is best captured in Marx's (2000: 72) view that religion is an expression of protest against injustice that provides consolation to believers in its shared expression. He also considered it an oppressive force – and a critical element of capitalist ideology – because it quelled the temptation to revolt against injustice by promising a divine form of justice in the afterlife. Due to its social nature, the strategy of eradicating religion through scientific education (the preferred approach of scientific atheists) was bound to fail, in Marx's view, because this did not address the social conditions that grant religion its appeal and its power. He believed that only a social revolution that overturned the class system would finally eliminate the desire for divine justice, the belief in God and the power of the church (Chadwick, 1975).

Humanistic atheism is therefore as much a political project as scientific atheism. Both feature not just a theory purporting to explain religious belief, but also a vision of what kind of society we would need to construct in order to get rid of it. For humanistic atheism, the answer is social justice and an end to inequality, while scientific atheism emphasizes the freedom and authority of science, which is assumed to drive social progress. These two forms of atheism are not antithetical – in fact, the atheists in the 'anthropological approach' took many of the premises of scientific atheism and Enlightenment criticism as a starting point – but, in practice, they have become poles in tension. This is particularly evident in the New Atheism and the contemporary atheist movement.[1]

## The New Atheism

The New Atheism is new only to the extent that the scale of its impact on public discourse was unprecedented. In terms of the arguments it makes and the ideology it promotes, it is really little different from the scientific atheism of the Victorian period, except that it is updated with theories derived from the nascent disciplines of evolutionary psychology and neuroscience. Drawing on these new sciences, the New Atheists portray religion as a pre-scientific explanation of nature – what Richard Dawkins (2006) refers to as the 'God hypothesis' – but also as a natural phenomenon. This means that religion is produced by natural forces rather than social forces, and these can be understood by recourse to evolutionary theory. The natural forces in question are biological – Dawkins and Dennett, in particular, have outlined a theory of religion as the outcome of genetically determined tendencies to attribute agency to inanimate objects and natural events. A closer examination of their critique reveals that they reproduce scientific atheism's ideological defence of the authority of science and its expert practitioners, and advance their own belief system as an alternative to religion.

The belief system in question involves epistemological, ethical and political prescriptions that outline a basis for ordering society according to expert knowledge. The most fundamental principle of this belief system is scientism, which is the view that our knowledge of reality is limited to that which science has access to, and that what lies outside of the reach of science cannot count as knowledge (Stenmark, 1997: 19). In other words, science defines the limits of knowledge, but, further, it defines the limits of reality. That is, if something is unknowable by science, then it cannot be said to really exist. By 'science', I mean the natural sciences, the boundaries of which are extended to include not just the natural world, but the social world, including human behaviour and institutions. In its more extreme forms, scientism extends the authority of the natural sciences into areas normally considered the province of philosophy, including morality (Olson, 2008).

Scientism, then, involves two major characteristics: a statement on the limits of knowledge and the nature of material reality; and an extension of the authority of the natural sciences, specifically, to non-natural or immaterial social and cultural phenomena. Combining these perspectives, I offer this synthesized definition – scientism is the view that: science is the only legitimate form of knowledge; that the domain of the natural sciences may be expanded to encompass human institutions, behaviour and value structures; and that the theories and methods of the natural sciences are the best approach to the study of society and culture.

The ideology of New Atheism is characterized by a specific form of scientism that I call evolutionism, which places explanatory authority at the feet of Darwinism, with evolution a general law that can be applied to living things and the universe as a whole, but also human society and culture. In their version of scientific atheism, evolution not only is a challenge to religious belief, but also serves as a theory of human history. The scientistic application of theories and concepts derived from Darwinian evolution to the social world is the basis of the social theory and political doctrine at the heart of the New Atheism, which is more a defence of Western modernity and its social organization and structures of power and authority than an intellectual critique of religious belief (Eagleton, 2009; Stahl, 2010; LeDrew, 2012).

Evolutionism as a social theory translates into a teleological notion of social progress, with all cultures advancing from superstition to scientific enlightenment, and from a power structure defined by religious authority to one defined by scientific authority. This is the basis of the New Atheism's celebration of the Enlightenment, and modernity generally, for establishing the social and cultural conditions where scientism can become entrenched as a universal world view (while it is not yet universal, and religion has proved more stubborn than expected, they believe that the triumph of science over religion is inevitable and imminent). Faith in progress is evident when Dawkins (2006: 270) suggests that there is a steady change in social consciousness in a relatively consistent direction in modern liberal-democratic societies. He does admit that there are challenges and interruptions to this progress, but nonetheless believes that progress is inevitable: 'Of course, the advance is not a smooth incline but a meandering sawtooth. There are local and temporary setbacks.... But over the longer timescale, the progressive trend is unmistakable and it will continue' (Dawkins, 2006: 271). For Dawkins, this social progress is driven by science, and societies that accept scientific authority invariably become better places.

While religion is the most direct threat to the New Atheism's ideology of scientism, and its chief antagonist, there is another rival form of thought that receives almost as much attention and fierce criticism from New Atheists: 'postmodernism'. While the term is rarely interrogated systematically, the New Atheists use it to refer primarily to relativism, but also use it to label people who advocate for 'social justice', and apply it to what they deem to be unscientific thinking in general. Epistemic relativism is an obvious challenge to the scientistic absolutism that the New Atheists endorse, and, further, it allegedly legitimates religion by undermining the validity of rational-scientific critique. In their view, relativism is embodied and legitimated by the social sciences and humanities, under the influence of the intellectual paradigm of postmodernism, which, in a signal of their ignorance of these fields, they assume is characteristic of all such scholarship. There is therefore an undercurrent of criticism in New Atheist thought regarding the exclusion of the social sciences from the domain of legitimate knowledge, as defined by scientism.

## Atheism and the science wars

The New Atheists are hostile towards the social sciences primarily because they consider those disciplines to support and promote relativism, and therefore represent a rival form of thought that must be excluded. However, they also see the social sciences as basically obsolete since we have the new Darwinistic disciplines of evolutionary psychology and neuroscience to perform the same function. The strategy is to argue that the social sciences are irrelevant because Darwinism provides superior explanations of human society and behaviour. This is the view advanced by Harvard entomologist Edward O. Wilson (1975), who famously argued that the social sciences should be considered an undeveloped branch of evolutionary biology, and should therefore be replaced with the new science of 'sociobiology'.

Richard Dawkins takes the same view, disapproving of the supposed fact that the social sciences and humanities 'are still taught almost as if Darwin had never lived' (Dawkins, 1989 [1976]: 1). Dawkins believes that the theory of natural selection has explanatory powers that reach well beyond the evolution of life. He claims that human culture evolves progressively in precisely the same way that species evolve, that is, by natural selection:

> Fashions in dress and diet, ceremonies and customs, art and architecture, engineering and technology, all evolve in historical time in a way that looks like highly speeded up genetic evolution, but has really nothing to do with genetic evolution. As in genetic evolution though, the change may be progressive. (Dawkins, 1989 [1976]: 190)

The difference is the unit of transmission: in biological evolution, it is the gene; in cultural evolution, the 'meme' (roughly analogous to 'idea') is the unit that is negatively or positively selected and transmitted. Dawkins declares that memes are the 'new replicators', transmitting culture from generation to generation just as genes transmit other physical and behavioural traits.

Dawkins's theory of religion is built on both evolutionary biology and his meme-based theory of culture, and involves the principle of biological predisposition to religious belief, coupled with memetic transmission as the basis of the growth of particular belief systems. The basis of religious belief, then, is a by-product of evolutionary adaptations: 'The religious behaviour may be a misfiring, an unfortunate by-product of an underlying psychological propensity which in other circumstances is, or once was, useful' (Dawkins, 2006: 174). This is most obvious in the case of children, who Dawkins claims are hard-wired by genetic evolution to trust the words of their elders, a useful adaptation for survival but with an unfortunate by-product being 'vulnerability to infection by mind viruses' (Dawkins, 2006: 176).

Dawkins considers explanations of religion provided by the social sciences to be 'proximate', and not 'ultimate', explanations. Darwinism, he argues, is an ultimate explanation that gets to the core of religious belief, and it involves an application of the theory to two things: our brains and the genetic evolution that gave rise to them; and ideas that exist outside of these brains in a meme-pool, which is also subject to the process of natural selection. He does deal very briefly with 'proximate' explanations, noting that 'Religion has at one time or another been thought to fill four main roles in human life: explanation, exhortation, consolation and inspiration' (Dawkins, 2006: 347), but he considers these minor and relatively insignificant factors.

The New Atheists' theory of religion, then, doubles as an argument for replacing the social sciences and humanities with Darwinism, which is an effort to legitimate the authority of science outside of its natural boundaries. This is reflected in the atheist movement more generally, where representatives of the natural sciences dominate discussion, and representatives from the social science of religion are almost completely absent (LeDrew, 2015). The atheist movement does not look to social science because of the presumed universal authority of the natural sciences. Rather than exploring the social dimensions of religious belief and practice, the New Atheists look to biology and evolutionary psychology to find causes of religious belief, which is understood as something that arises from our material bodies and the evolutionary processes that have shaped our brains and our thinking. The nature of religion, then, is understood as a product of biological forces that have shaped human nature, with Dawkins's theory of memes added to explain how religion takes shape in different sociocultural contexts – or, in his terms, how memes replicate and produce different beliefs in different 'environments'.

Daniel Dennett is even more explicit about his support for Darwinistic interpretations of sociocultural phenomena and his portrayal of the social sciences as an antagonist to scientific truth. This is clear when he writes:

> Anyone who tries to bring an evolutionary perspective to bear on any item of human culture, not just religion, can expect rebuffs ranging from howls of outrage to haughty dismissal from the literary, historical, and cultural experts in the humanities and social sciences. (Dennett, 2006: 259)

Like Dawkins, he dismisses the 'proximate' explanations of religious belief, which are symptoms of a 'disorder often encountered in the humanities and social sciences: premature curiosity satisfaction' (Dennett, 2006: 103). This 'disorder' is demonstrated when questions regarding origins are 'left unexamined by people who lose interest once they have found a purpose or function for religion that strikes them as plausible' (Dennett, 2006: 102). The purposes that Dennett cites are comfort, explanation and cooperation, and with that he perfunctorily summarizes, and dismisses, the social-scientific understanding of religion. Instead, Dennett supports Dawkins's evolutionistic theory of religion, explaining its general nature as a by product of traits that were suitable to our ancestors' survival, most importantly, the tendency to believe that there is a conscious agent in all objects and natural events. Also like Dawkins, Dennett believes that Darwinism is sufficient to explain social and cultural phenomena more generally, claiming that 'in principle the process of natural selection is substrate-neutral' and that evolution will occur wherever the conditions of replication, variation and differential fitness (or competition) are met (Dennett, 2006: 341). In organic evolution, the substrate for natural selection is the unit of the gene, organism or group, and in cultural evolution, the substrate is ideas.

The New Atheism, certainly as represented by Dawkins and Dennett, might be understood as a manifestation of a resurgence of sociobiology in recent decades, though these authors prefer the term 'evolutionary psychology', which does not carry the stigma now commonly associated with sociobiology. Darwinists like Dawkins and Dennett who invoke evolution by natural selection as a universal explanatory framework effectively challenge the foundational assumption that the social sciences were built on, specifically, the rejection of biological determinism in the understanding of human behaviour (Bowler, 2003: 360). In essence, then, the project of sociobiology is to make the social sciences unnecessary (Rose and Rose, 2010: 112). Sociobiology is a foundational element in the project of scientific hegemony, demanding that only biological causes of behaviour should be taken into account, with humans as machines that operate according to genetic instructions, just like other animals and plants (Midgley, 1985: 151). It thus seeks to make the social sciences redundant, advancing the new disciplines of evolutionary psychology and neuroscience as superior alternatives.

The salient point here is that the rejection of the social sciences is more precisely a rejection of relativism. Replacing the social sciences with Darwinism is, in fact, an endorsement of absolutism, with the natural sciences providing 'ultimate' knowledge with respect to questions and issues that, from a sociological standpoint, have no clear answer and are not reducible to a single all-encompassing explanation. The relativism attributed to the social sciences is an obstacle to the scientific critique of religion, and, more importantly, the overarching evolutionistic vision that New Atheist discourse is grounded in. The social sciences are therefore reduced to an undeveloped branch of evolutionary biology, subsumed to what Dawkins (2006) considers the 'ultimate' theory of natural selection, which Daniel Dennett (1995: 82) views as a theory of such vast scope that it transcends disciplinary boundaries, 'promising to unite and explain just about everything in one magnificent vision'.

With its reductionist critique of the social sciences as relativistic and corrupted by 'postmodernism' (a term that is never precisely defined, but is applied loosely to any ideas that are deemed to be unscientific), the New Atheism echoes the 'science wars' of the 1990s, which saw a number of prominent scholars from the natural sciences, on one side, and the social sciences and humanities, on the other, engaged in a prolonged debate about truth and knowledge (Ross, 1996). The science wars were launched by the infamous 'Sokal hoax', where mathematician-physicist Alan Sokal published a sham article in the journal *Social Text* that was intended to demonstrate that cultural studies was a field filled with nonsense. Sokal followed up the prank with the book *Fashionable nonsense*, co-authored with Jean Bricmont (Sokal and Bricmont, 1998), which was an extended critique of the 'postmodern' social sciences and humanities that charged them with relativism and obscurantism.

While some of the criticism is merited, Sokal chose a handful of the most uncompromising and radical intellectuals (mostly French) as representative of a vast field of scholarship. Nonetheless, the hoax and the book were enthusiastically promoted by many scientists, including Dawkins, who actually wrote a promotional blurb for the cover of *Fashionable nonsense* that said 'the hoax was earnestly needed and richly justified'. Dawkins does not hide his contempt for the social sciences and humanities, particularly sociology, which he equates with relativism and a general distortion of scientific truth. His disdain for the discipline has come out most forcefully on social media, where he tends to express his views most frankly. In a 24 May 2013 tweet in response to accusations of racism for comments he made about Muslim 'barbarism', Dawkins wrote: 'So many people incapable of drawing an elementary distinction: between racism and INSTITUTIONAL racism. Probably studied sociology'. On the same date, he re-tweeted a comment made by one of his followers, a sign of a view he supports: '@RichardDawkins be fair, sociology allows McDonald's to get a slightly more educated staff pool'. When another Twitter user made the obviously problematic claim that one cannot by definition be racist or sexist against white men because they are the group holding power, Dawkins responded with: 'Really? By whose dictionary? Certainly not the Oxford Dictionary. Dictionary of sociology perhaps?'.

Other leading New Atheists have echoed these criticisms of sociology and philosophy, in particular. Lawrence Krauss, a physicist at Arizona State University and prominent figure in organized atheism (his speaking tour with Dawkins was the subject of a documentary film, *The unbelievers*), gave an interview to *The Atlantic* in 2012 that made waves primarily for his comment about 'moronic philosophers' who criticized his book *A universe from nothing* for its deliberately misleading title, which suggested that the creation of the cosmos was possible without a divine creator. (In the interview, Krauss was pressed by his interviewer to admit that, in fact, the evidence offered in the book did not indicate that the universe came from nothing.) Krauss's comment was directed precisely at the philosopher David Albert, who had panned the book in the *New York Times* (Albert, 2012). Krauss countered that philosophers contribute nothing to the advancement of science – a telling comment since philosophers are not scientists, reflecting Krauss's belief that science is the only kind of knowledge or intellectual pursuit of any value.

At roughly the same time, Steven Pinker, an evolutionary psychologist at Harvard University and another intellectual venerated by the New Atheists and their followers, wrote an article in defence of scientism in *New Republic* titled 'Science is not your enemy'. On the surface, it implored those in the social sciences and humanities to recognize their common interests with science and try to work together, while overtly ridiculing those same disciplines and arguing that science is the only legitimate source of knowledge. (Pinker's contempt is clear in the article's byline, 'An impassioned plea to neglected novelists, embattled professors, and tenure-less historians'.) Neil deGrasse Tyson, a physicist and host of the television mini-series *Cosmos* (an inferior remake of the Carl Sagan classic), created a stir in an interview promoting *Cosmos* when he described philosophy as a 'useless' discipline that has nothing to tell us about the natural world. Tyson echoed Krauss's misunderstanding of what philosophy is, seeing it much like religion: as a competitor in the quest for a rigorous understanding of nature, and a threat to the authority of science. While these attacks primarily centre on philosophy, the territorialism extends to the social sciences as well, particularly among Darwinists who advocate for evolutionary psychology and sociobiology as the basis for a new science of society.

The science wars have, then, not ended – at least not for atheists. In 2017, another Sokal-style hoax was carried out by some prominent atheists, this time directed at the field of gender studies. Philosopher Peter Boghossian and mathematician James Lindsay co-authored an article entitled 'The conceptual penis as a social construct', which was published in the journal *Cogent Social Sciences* (CSS). The article was about as nonsensical as Sokal's, though *Social Text* was and is a prominent journal, while CSS was virtually unheard of and, worse, is pay-to-publish, which essentially means that anyone who is willing to pay to have their work published will find a home in that journal. While the authors claimed that the article proved that the entire field of gender studies is completely useless, it was not long before the critics began pointing out that all they had accomplished was having a bad article published in a bad journal by paying them to publish it. Nonetheless, many New Atheists – Dawkins included – took it as further evidence that the social sciences and humanities as a whole are nothing but fashionable nonsense.

## Conclusion

The New Atheism's antipathy towards the social sciences is rooted in both epistemological and political concerns. The ideology of scientism, and more specifically evolutionism, which underwrites all New Atheist thought insists that the natural sciences are the only legitimate source of knowledge in virtually all spheres of inquiry. Their commitment to the epistemic supremacy of science and its boundless explanatory range naturally includes the view that the social sciences are an unnecessary addition to what can be known about human beings and their social world by looking through the lens of Darwinism. Scientism is a universalism that makes no space for other kinds of knowledge.

The political dimension of the New Atheism involves a defence of Western modernity and their particular conception of liberalism, grounded in the view – again, guided by evolutionism – that social progress is inevitable and driven by science. The social sciences, equated with relativism and postmodernism, are perceived as a threat to progress precisely because they undermine epistemic and moral universalism, allowing for the possibility that different cultures have equally valid grounds for their particular values and ways of life. This view, or a variant of it, is the basis of the policy of multiculturalism that is now embraced (though not without contest, clearly) by most Western societies, to one extent or another. While based on a liberal political philosophy, multiculturalism is antithetical to the universalistic world view of New Atheism, which offers a vision of a world where law, morality and justice are determined and administered by scientists, who are rightly invested with authority.

## Note

1 See also Kind (Chapter 9, this volume), which delineates this tension within the contemporary Swedish secular humanist movement.

## References

Albert, D. (2012) 'On the origin of everything: "a universe from nothing", by Lawrence M. Krauss', *The New York Times*, 23 March. Available at: www.nytimes.com/2012/03/25/books/review/a-universe-from-nothing-by-lawrence-m-krauss.html (accessed 20 March 2018).

Berman, D. (1988) *A history of atheism in Britain: From Hobbes to Russell*, London: Croom Helm.

Bowler, P.J. (2003) *Evolution: The history of an idea*, Berkeley, CA: University of California Press.

Buckley, M.J. (1987) *At the origins of modern atheism*, New Haven, CT: Yale University Press.

Buckley, M.J. (2004) *Denying and disclosing God: The ambiguous progress of modern atheism*, New Haven, CT: Yale University Press.

Chadwick, O. (1975) *The secularization of the European mind in the 19th century*, Cambridge: Cambridge University Press.

Dawkins, R. (1989 [1976]) *The selfish gene*, Oxford: Oxford University Press.

Dawkins, R. (2006) *The god delusion*, Boston, MA: Houghton Mifflin Company.

Dennett, D. (1995) *Darwin's dangerous idea: Evolution and the meanings of life*, New York, NY: Simon and Schuster.

Dennett, D. (2006) *Breaking the spell: Religion as a natural phenomenon*, New York, NY: Penguin Books.

Eagleton, T. (2009) *Reason, faith and revolution: Reflections on the God debate*, New Haven, CT: Yale University Press.

Harris, S. (2004) *The end of faith: Religion, terror, and the future of reason*, New York, NY: W.W. Norton and Company.

Hyman, G. (2007) 'Atheism in modern history', in M. Martin (ed) *The Cambridge companion to atheism*, Cambridge: Cambridge University Press, pp 27–46.

LeDrew, S. (2012) 'The evolution of atheism: scientific and humanistic approaches', *History of the Human Sciences*, 25(3): 70–87.

LeDrew, S. (2015) *The evolution of atheism: The politics of a modern movement*, New York, NY: Oxford University Press.

Lightman, B. (2001) 'Victorian sciences and religions: discordant harmonies', *Osiris*, 16(1): 343–66.

Marx, K. (2000) *Karl Marx: Selected writings* (2nd edn) (ed D. McLellan), Oxford: Oxford University Press.

Midgley, M. (1985) *Evolution as a religion: Strange hopes and stranger fears*, London: Routledge.

Olson, R.G. (2008) *Science and scientism in nineteenth-century Europe*, Urbana, IL: University of Illinois Press.

Rose, H. and Rose, S. (2010) 'Darwin and after', *New Left Review*, 63(May/June): 91–113.

Ross, A. (1996) *Science wars*, Durham: Duke University Press.

Sokal, A. and Bricmont, J. (1998) *Fashionable nonsense: Postmodern intellectuals' abuse of science*, New York, NY: Picador.

Stahl, W.A. (2010) 'One-dimensional rage: the social epistemology of the New Atheism and fundamentalism', in A. Amarasingam (ed) *Religion and the New Atheism: A critical appraisal*, Leiden: Brill, pp 97–108.

Stenmark, M. (1997) 'What is scientism?', *Religious Studies*, 33(1): 15–32.

Whitmarsh, T. (2015) *Battling the gods: Atheism in the ancient world*, New York, NY: Knopf.

Wilson, E.O. (1975) *Sociobiology: The new synthesis*, Cambridge, MA: Belknap Press of Harvard University Press.

PART IV

# Religion, Conflict and Moderation

11

# Science and the Unearned Virtues of the 'Really Religious People': Exploring the Association between Perceived Religiosity and Science Rejection among Students in the Midwestern United States

*David E. Long*

## Introduction

In many cases, a person's religious identity neither conforms completely to the tidy labels that social scientists use to categorize them, nor is an adequate model to capture the reasoning that a person uses to maintain that identity. As Wuthnow (2008) has described with regards to Americans' faith practices, our identities look more akin to inclusionary or exclusionary stylistic types rather than to closely following specific tenets. Someone who describes themselves as a member of x or y faith community has much in common with others within that community as an affinity group, but what this is typically varies quite a bit from person to person in terms of both the specific knowledge and ideological orientation of their faith. While this may have differed in the past, recent analyses of the US public have shown that 'large numbers of Americans are uninformed about the tenets, practices, history and leading figures of major faith traditions

– including their own' (Pew Research Center, 2010). People share nominal identity, but upon scratching them for the specifics, they often express limited or superficial understanding.

In the US, a long history of supposed conflict between religious faith and scientific knowledge has marked the relations of certain politically and socially conservative Christian denominations towards parts of science (Numbers, 2007). Since its arrival in the US, evolutionary theory has posed cognitive and social problems in the minds of certain believers for whom the story that sustains their religious identity is at odds with, and must be rejected in order to accept, evolution. American Christians have dominated this story, but it also includes small minorities of Muslims and Jews, which draw less attention due to small populations of the latter groups in the US (Guhin, 2016). While Christian religious identity has waned slightly in recent years (Pew Forum on Religion and Public Life and Pew Research Center, 2008), the types of believers with a sense of potential conflict with evolutionary theory may be a larger proportion of Christians within an overall slightly shrinking population (Schnabel and Bock, 2017).

In such a context, what can we say about those Americans who either seek to reconcile points of conflict between science and faith or, by virtue of their existing practices, rarely or never experience such conflict? How do they account for, and think alongside, those for whom evolutionary theory poses cognitive and interpretive challenges, and, at times, an entree to existential crisis?[1] Why do, as we shall see, such ecumenically minded Christians attribute the status of 'very religious' upon those they see as rejecting evolution? Why, in their categorical imagination, does their own ecumenist position exist in a less charged, or 'extreme', position in relation to those they deem as 'very religious'? Why, historically, is this the case, how does this compare internationally and what would it mean to look at the virtues cultivated by faith practice in different ways? What would it mean to detach our understanding of religiousness from one where religious conservativism as seen today is no longer the end of a supposed scalar relationship, towards one of more historically aware styles of practice? For some researchers (eg Allum et al, 2014), evolution rejection is a proxy for strong religiosity. What I am suggesting is that we may be better served by moving away from such linkages as they give undue credence to a small segment of the social field. One could, for example, be a very religious Unitarian, a very religious Presbyterian or even a very religious Lutheran; none, as the sociological imagination is currently conjured, would qualify.

While these questions are much larger than can be answered here, by looking through the eyes of Americans and their relationship towards religious difference, we at least see pathways for differing expressions of faith where science does not pose such intractable conflict – although recent changes in the demography of US religious identification pose challenges to this hope, which must be thoughtfully accounted for.

## Context

Sociologists such at Robert Wuthnow (2008) and Anne Swidler (1986) have provided useful frameworks for understanding the means by which people practise and hold to their identity, and the conditions and direction by which they experience changes of identity when they take place. Wuthnow gives us the categories of the religious 'exclusivist', the person who knows that he or she possesses the one true way. Those who take an ecumenical or other more socially accommodating approach, seeing both their faith and others as viable interpretive frameworks for understanding the world, he describes as 'inclusivist'. Finally, those inclined to drift from practice to practice in an almost polytheist manner – not deep believers in any one tradition, but rather following myriad styles of practice situationally and conditionally as times and fashions change – he terms 'spiritual shoppers'. As I use in my work (Long, 2011), the logic of Swidler's (1986) 'strategies of action' gives us a pathway to understand the seeming non-rationality of religious believers when belief becomes more difficult, well before the current attribution of non-rational action as 'motivated reasoning' (Kahan, 2013).

In my own ethnographic work in Middle America among university students and their extended social networks, sustaining one's religious identity when it potentially conflicted with evolutionary theory saw people experiencing an aversion to the topic based in existential anxiety, with strong cases experiencing emotions indicative of existential crisis when asked to explain the dissonance between their narrative of faith and that of science. Such students typically adhered to an evangelical or Pentecostal identity, but not always. Some were self- or other-identified as 'conservative Catholics'. If people of these supposed 'extreme' views – the 'very religious', as many termed them – experienced changes of faith, it was not a change as matter of degree. It was all or nothing: fundamentalism to agnosticism or atheism, and vice versa. The seeming moderate 'both/and' position of Wuthnow was not looked upon as a viable middle way option. Why is this?

Like many matters of identity, cultural affinity is learned locally, within kinship networks and in distinctive communities, and is complicated. Being a member of a religious group, in the US at least, but certainly not limited to this research, is as much or more about a closely held and experienced communal sense of identity – a liturgical one – than necessarily adherence to a deep understanding of the textual theology and its implications for action (McGuire, 2008). People often have deeply felt convictions about their religious identity, with often shallow or contradictory understandings of what it means to be the thing they represent (Bloom, 1993). Self-aware reflection on what constitutes their identity is not a comfortable or natural place for most people to dwell in.

Untreated in my ethnographic work with American college students grappling with the interface of science and faith during their university undergraduate studies was a curious phenomenon that begs further analysis. Aligned with questions asked by Michael S. Evans (2016), how and why do we, as scholars and communicators interested in religion, collaboratively construct and endorse an interpretive framework that places those who adhere to Biblical inerrancy on a supposed far side of all American faith traditions? How and why does this become the 'extreme' side of a sliding measure? Why is this? What implications does this have for thinking through the interfaces of science and faith in conflict versus other social imaginaries and or realpolitik solutions to, or prescriptions for, alternative social worlds to be constructed?

To work with these questions, interviews with university students who made up the core of a 2011 study (Long, 2011) serve as a touchstone by which we can examine bigger questions. I am interested in how certain styles of religious practice come to dominate the imagined 'extreme'. How do they become the supposed terminal end of a scalar measure of religious intensity?

Such imagined ends, I contend, are neither accurate nor inevitable. In this way, discourses from the coverage of Islam in broadcast media that created an identity of 'the moderate Muslim' are useful. Just as violence attributed to religious extremist actions in the name of Islam does not represent the most legitimate form of the faith's practice, but rather that of a small minority, so neither does creationist disagreement with scientific orthodoxy. For example, in this view, the type of Christian who espouses creationist views of the world is held as a terminal marker upon an imagined linear scale of religious intensity and, in some cases, legitimacy. While, in any once slice of time, this is a functionally useful tool, such distinction misses the implications from the longer, ethnological story of science, faith and interpretation. Many American Christians, for example, have no idea that creationism, as they know it, is a relatively recent phenomenon when measured against all of Christian history.

In classes that I teach on the history and philosophy of mathematics and science, American students are often puzzled at the idea of unsettling the surety of their theological 'tribal' identity for the purposes of social and historical analysis. Most people are simply not primed to see much opportunity or reason to do so. Creationism, in the most commonly invoked US examples, is a fairly recent historical phenomenon when one considers the over two millennia story of Judaeo-Christian religion and the more recent advent of contemporary science and initially correlated phenomena. Creationism, as we understand it today, only becomes a possibility in a social milieu where a generative hermeneutic tradition of religious interpretation has emerged in response to an unfolding scientific story of human origins. Exacerbated during the 19th century's revolutionary thinking about the biological mechanism of species change via evolutionary theory, we in the Judaeo-Christian-influenced West have been dealing with social forces shifting the dominant interpretive framework away from Christianity's epistemology and ontology. During this process, Christian denominations ascribing to 'mainline' identity began to split markedly with denominations that would give rise to today's evangelicals and Pentecostals – those most likely to have quarrel with evolutionary theory (Marsden, 2006).

Thus, of course, for anyone wishing to situate a stable narrative that makes up the core of one's identity, disruptive stories that confound that identity are not welcome. In order for ideology to remain stable, social forces that threaten change are marginalized (Swidler, 1986). In the case of students in classrooms where ideologically divisive territory is to be tread, we *should* expect difficulty. The coarse cultural milieu of varied religious and epistemological commitments spilling into the discourse of the public square should, in a national context that has worked so hard to create a form of 'public niceness' (Varenne, 1990), be alienating. We should expect mechanisms to suppress possible social tensions over such differences to have long been developed and refined. Indeed, as Chan and Ecklund (2016) have found, distinctly different strategies of epistemology management among differing Christians seemed to depend on the specific instances of culture and conflict, not broad national consensus or trends.

What we are after here, then, is twofold. What does close examination of ecumenical Christians' thinking about science and faith, particularly the 'Other' of creationist Christians, and their projection by ecumenicists into an 'extreme' position of 'very religious', tell us about the scalar sense of religious identity in the US? Echoing perspectives raised by O'Brien and Noy (2015), it appears more accurate to describe some Americans' relations towards faith and science through a dominant lens of culture and epistemology – not simply that of opposition. This sense becomes important as we question why, at least from the perspective of Aristotelian virtue ethics (MacIntyre, 2010), such 'very' religious positions gain their spot on an imagined scalar relation. Is this a socially desirable state of affairs? Is this imaginary doing advantageous work for those who look, as science and religion conflict entrepreneurs, to exacerbate tensions that might otherwise be smoothed over? With this thought through, can we imagine alternatives that might serve a public or formal educational good for advancing both science and/or faith?

## Method

My investigation of these questions includes an analysis of interviews conducted with American university students at Mason-Dixon University.[2] Mason-Dixon is a regional comprehensive public university selected for its proximity to a major creationist organization, but also for its indicativeness of 'average' US college-going tendencies. Regional comprehensive universities graduate the largest number of degree-holders in the US by far, while Mason-Dixon is specifically situated geographically on a border that has long typified cultural and political identity boundaries between the US South and Midwest – two of the historically most prominent regions for science and religion tension. While the respondents in question are university students ranging from 19 to 26, I am not making a claim that this population is completely representative of all faith development across a lifespan, or the scientific literacy that one might expect to be gained while at university. These data and analysis are a continuing story begun in a larger ethnographic study (Long, 2011), and are used as a specific data set to open new angles of questioning.

The motivation to bring this analysis to light has much to do with the rise of the 'nones' religious identity generated within the past two Pew Research Center religious landscape surveys (Pew Forum on Religion and Public Life and Pew Research Center, 2008, 2014). Whereas the 'both/and'-minded inclusivist faithful that Wuthnow (2008) describes are a convenient and familiar way to talk about mostly science-positive folks of historically non-literalist faiths, the 'nones' are a unique recent categorization choice by Pew. With neither identification nor deep history with religious practice, these people are not the agnostics or atheists of earlier iterations of social science research. Religious identification is apparently simply something they do not do. Importantly for historical categories of science and religion studies, these 'nones' may be no more or less inclined to even care about science. Queering the earlier category, the absence of religious practice may not have any effect on interest in science. Importantly for marking these data in time, the inclusivists with whom I spoke and focus on here may be the trailing edge of their faith's social cohort, being replaced now – eight years since these data were collected – by a growing number of 'nones' today.

Students at my research site were selected to represent examples, analytically, of the religious 'inclusivist' identity category following Wuthnow (2008) by virtue of their expressed ecumenism in the content of their discussions. These were students that, as a group, never brought forth critique of faith as a limiting factor in their interpretive toolbox for examining the world – it always aligned at least complimentarily with faith. Expressed ecumenism was categorized as representing those students for whom science and faith were valued parts of their identity, and who looked to both to support part of their identity, either in tandem or as a interwoven whole. This ecumenism, for Wuthnow, was a 'both/and' position. Both science and faith would work in parallel or harmoniously in the interpretive framework of their lives. The broader research context and framework can be examined in Long (2011). This earlier work focused more on the motivated reasoning of creationists as they went through the university experience and thought about evolutionary theory working within the framework of their lifeworld and epistemology. Of note, especially in light of recent US political polarization and the geographic isolation of disparate ideological communities, familiarity with and understanding of such religious interpretive difference seems all the more needed.

Religious inclusivists become interesting in the context of changing US religious identity as they give us an example by which we can explore the public sphere of dialogue – the exemplar being schools – and how this is seemingly more and more at risk by changing demographics and the intensifying ideological framing of almost every issue. As I will examine first: what is the content and descriptive understanding of the religious inclusivist towards that of the religious exclusivist? How well do they know the content of their own faith and that expressed by the Other? In what way does this type of understanding prepare those that carry it to engage in the broader political work of negotiating science and faith in the public sphere?

This comes to matter as, I will suggest, we are standing at the edge of an opportunity. Religious inclusivity is equipped to broaden and support religious difference across a diverse population, whereas religious exclusivity does not, if left to its own ends, tolerate religious difference – it aims to negate it. Such inclusive potential is at risk as US religious demographics shift away from tolerating difference towards a monological vision of exclusivism. Addressing this will mean reconfiguring certain conceptual schemata that have guided conceptions of how science and religion interact with each other.

Continuing, I will examine alternative models of science and faith interaction looking forward, using ideas gleaned from Taylor's (2007) understanding of the pre-modern, modern and post-modern condition. The post-modern condition and, specifically, typical science and religion 'debates' become much more understandable when cast in this frame. At least in the US, when Anabaptists such as the Old Order Amish, scientisitic scientists such as Richard Dawkins or Neil DeGrasse Tyson, and postmodernists such as Judith Butler or Edward Said interplay in the public sphere, they are all coming from very different starting points and often occupying quite different geographies. Contemporary discourses of science and faith mostly ignore this complexity as teasing apart such complexity has largely been the work of anthropologists (to understand cultural difference) and philosophers (to understand the grounding of claims based upon varying epistemologies) (Butler et al, 2011). What I point towards is a new view towards framing religious discourse vis-a-vis science as a form of Aristotelian virtue ethics, albeit cut off from ultimate grounding. Diverging slightly from MacIntyre (2010), we have no practical choice in disconnecting from ultimate grounding for the health and maintenance of an open, pluralist society. This, in turn, sees the public square of civic discourse turn towards an underpinning of utilitarian goods, not a stance assuming means towards ultimate ends.

## Religiously inclusive Americans' imaginations of the 'very religious'

Historicist theologians such as Borg (2006) have pointed out the erosion of US mainline Christianity for the past 30 years. Studies of the diversity of American religious identity and practice by the Pew Research Center have corroborated this. While Americans are slightly less religious on the whole than they were 30 years ago, of those who believe, the percentage of believers of mainline Christian identity versus evangelical or Pentecostal identity has quickly changed. Mainline Christianity is shrinking while evangelical and Pentecostal Christians are quickly coming to dominate the numbers of those who believe. This comes to matter as the social skill set of ecumenism cultivated within mainline churches is anathema to evangelicals and Pentecostals. Christian ecumenical social efforts of the 1970s and after are simply not shared ideologically by the religiously conservative (Hunter, 1991).

If social change regarding faith traditions depends on outward trust and camaraderie between faiths, then evangelicals and Pentecostals are unlikely to drive such change. Of the ecumenically minded students I spoke with, what are their hopes regarding ameliorating social discord over science and faith? The short version of this is very little as both the ecumenically minded and religious conservatives are aiming at differing ends of ontologically different worlds. One sees pluralistic harmony as a social good, while the other sees one true light to which all others must align or be eliminated.

As we consider the ongoing tension between science and faith, and whether such supposed tension can easily be worked past, how do ecumenically minded students view the religious Other? The following sections describe an analysis of 16 students' interviews concerning their understanding of religious diversity, and how they view those who reject evolution on religious grounds. I spoke with these students using a common interview protocol that was designed to elicit the breadth and extent of their understanding of their own faith, and how it related towards differences they see in other denominations' faith. Across these interviews, five distinct areas of discourse emerged relating to their understanding of their own faith in relation to religious exclusivists as played out in schools, moving from the general, through the political, to the local: science and faith; politics and the limits of creationist thinking; science, evidence and social legitimacy; science in public; and science and religion in schools.

## *Science and faith*

These students were distinctive in the manner in which they held what liberal theologians would term an ecumenical view of religious difference. For the most part, they found a way to balance the commitments of their faith with an understanding of the findings of science. Heather describes a typical interaction considering the status of evolution and her faith: "Once in catechism – my family attended a Lutheran church. And I'd talk about it with my Dad. He was in a science field, but just because he's a scientist doesn't mean that he doesn't have faith".

Distinctive among the religious believers within the broader study, a positive relationship between science and faith appeared to be actively cultivated by this ecumenically minded subset. Megan describes the style of this discourse:

> "Evolution is always being discussed in my family. They do see the scientific evidence and they don't try to say it's not valid. A lot of times, if I hear or see something interesting in class, I would discuss it at home."

At the same time, many of the students described social settings or styles of interaction where a clear dichotomy was presented to them; they just didn't adhere to this kind of distinction themselves. As John states, the conventional metaphysically naturalistic approach to science was well within their sphere of experience: "Basically, from what I've been told, it's like the world just coming into being on its own. That no God did it – that things just came about on their own".

As was the case in my earlier research on science and religion (Long, 2011), Catholics or ex-Catholics were interestingly split on how they perceived the right relationship for Roman Catholics towards science. Some perceived that the devout Roman Catholic's view was to clearly delineate science from faith, often taking a more conventionally creationist view of science than that of the Vatican post-Pius XII.[3] As Christopher details, their relationship towards their faith's dealing with scientific knowledge has changed:

> "I was born Catholic and all that. I was told that God did all this stuff, but as I got older, I don't really buy into that anymore. While it kind of makes sense, I kind of lean toward evolution and maybe it is true that everything just happened."

Similarly, Stephanie expresses the same kind of perceived Catholic 'conservative' intensity from her father: 'My Dad is Catholic as can be … if I would say something about evolution, I know he'd say I'm wrong". As she details, her mother, unlike her father, is not an active church attendee.

Other Catholic students, while expressing an ecumenical approach to matters of science and faith generally, still seem to harbour a keen sense that evolution is problematic for some. Asking Stephanie whether evolution poses a challenge to her religious identity (after expressing some tension), she describes the nature of this tension:

> "It certainly does create tension. My religion believes that a higher power created life as it exists today. Of course, evolution explains life – how it evolved from one stage to another. The idea in the eyes of a religious person, the kind I used to be, is that it takes away the divinity of the creation. So, it's not a popular topic with Catholics."

Additionally, Mark shows how exacerbated this perceived tension can get from the perspective of a non-Catholic. Reflecting on what constitutes 'extremist' views, he speaks generally of others' faith and who specifically has issues with evolution from the perspective of their faith: "For sure, other religions, Islamic religions and extreme Christians, will always combat the theory of evolution".

These more ecumenically minded students, while still carrying with them some tension regarding science and faith, were more positive about their outlook for their own future. Acknowledging a more historically nuanced understanding of the issue, William explains that "[E]volution has a history of being controversial between people about scientific and religious issues and how life began". Charlotte goes on to explain that their own approach is to actively look to find points of harmony and not discord. Typical of many of these students, Megan explains that her family's ecumenism had found a way to avert tension: "My parents are pretty at ease with balancing the idea of evolution with the idea of creation". As she goes on to explain, their parents had gone to great lengths to teach them that religious interpretation is just that – and may need to change with the changing times. Further into Thomas's interview, he is clear about the implications of this for his own future children: "I just want my kids to be able to think. As a father, you get all the information – you get everybody's sides – then you decide what to do". Catherine described the most coolly rationalist and somewhat social Darwinist perspective that I encountered among this group as she described the implication that she saw in the semantics of the word evolution: "It means change … to get better. It means change for the better. I mean like natural selection or whatever. I thinks it gives people … it makes them stronger and makes them strive for things in the future".

Extending this, some students who had been raised with or adopted a both/and position delineate that conflicts sometimes occur between individuals or among families. As Juan relates: "[T]hese issues can create tension, but maybe because of their personal beliefs. Their affiliation might actually be okay with evolution". Juan echoed something that I and many others have found: nominal religious identity is often fiercely tribal; however, the specific content or extent of that knowledge may be highly variable among members of that same close-knit group.

### *Politics and the limits of creationist thinking*

At the same time that inclusivist students expressed how science and faith related in practical terms to their own lives, they also projected what kinds of relations they imagined others to have. For these students who identified as finding both science and religion to not be in conflict in their lives, they also had experience of interacting with others who did not make efforts at accommodating both, or saw them as competing and a need for an epistemological 'winner'. Not all the inclusivists that I spoke with had a clear way to talk about these relations, but some had the sense that there were political and/or religious conservatives who, by virtue of how they talked about this generalized 'Other', were placed at the end of a projected spectrum of religious identification. Timothy begins to tease out the way in which these students thought about this kind of religious 'extremism':

> "People think that just because you think about human origins, their first thought is like – how could we come from monkeys? They don't really think that by saying we are related to monkeys that it's like, 'poof', monkeys to man. And it's like they don't get that it's the case that we slowly evolved. They just, they take the Bible literally, just word-for-word."

Some students expressed a more pointed experience of trying to negotiate science and faith conversations with those they saw as conservative. Stephanie described interactions with her friend that were clearly charged: "I have a friend who says that 'evolution is a load of crap'.... [S]he believes that the Bible is the word of God and that's it". Others furthered their analysis of the scenario, adding a layer of political identity: "The problem is that it's basically those that are right-wing conservatives, and take the Bible as being black and white, and there is no room for interpretation", Thomas said.

Past the individual evaluation of others' thoughts or expressions, yet other students extended this kind of perceived tension to the imagined social interactions perceived to be possible within their social networks. Megan sized up how her imagined conflict might arise and why:

> "If someone's like real religious … I mean if they're trying to say that the science trumps their religious views, I mean, that could start something. The only way I think it's a problem is if people are put head to head arguing."

Many, but not all, of the students I spoke with had an understanding that their own religious views were in some way more 'moderate' than those they saw as polarized or extreme. They also saw that some people "have a more theologically conservative view of their own faith than even their faith leaders see it", as Brittany put it. However, importantly, a key element of this identity was also their adherence to balancing science and faith, not looking to abandon one or the other. As Thomas observed: "I'm just open … I'm not an atheist because I do believe in a God". For these students, when looking at their identity vis-a-vis a perceived tension between science and faith, they took what they saw to be a moderate position, holding onto both as elemental parts of their sociocultural identity.

## *Science, evidence and social legitimacy*

Across the interviews, one distinctive aspect of these students' approach to science and faith was this 'both/and' ecumenism about varied approaches to interpreting the world, and their use of each as a non-competing toolset. This is not to say that potential conflict was never apparent to these folks; rather, it is that it was seen as socially undesirable as a communication or social organizational tool. By and large, my impression was that, above all else, through their experiences in their families, with most peers and within their church communities, they valued the ability to 'go along to get along'.

Some of the students expressed vague pleasantries to make this point: "We should take the views of every kind of people, and every kind of culture, into account. Everybody changes their views over time, just at different rates", shared Mary. Equally vague concerning dialogue at the interface of science and faith, Michael would "be fine with it … it's just opening them to more ideas and what others think". Many of the respondents saw such liberally minded talk as generally positive, as Stephanie explains: "I would not be opposed to it.… I think both theories are important to the educational process … I think providing options is the best way to do it". This is not a small point: the discursive move of attributing science and faith as competing theories, even if an artefact of a blunt set of dialogical tools, is a telling marker for later conversation. As Walter explains, even though these students were ecumenical as a categorical identity, their categorical dialogue is still shaped by social imaginaries outside their control:

> "There's some parts that I just kind of question … that I'm not sure. When we were talking about [science and religion] sometimes, my teacher made me think really hard. Just the whole thing – sometimes I thought that maybe some of it had to be true but maybe not other parts."

This kind of mental rumination was more pointed for some students. Megan remarks:

> "I know for certain that I believe in God. I've read *Mere Christianity* by C.S. Lewis, and *The case for Christ*. Who's to say that God didn't take all that time to create everything? I'm still learning about my faith."

Still thoughtful but a bit more hesitant, Jonathan explains why he does not have definite standpoints on many of the points of conflict between science and faith: "I think because of my beliefs…. I think that some of this might not be right".

While a clear minority, some of these ecumenically minded students were more enthusiastic about the power of science in their interpretive repertoire to explain the world: "You can see with your own eyes bone structures that are similar and have evolved from similar plans", Elizabeth explains, "[T]hey have phylogenetic trees now and understandings of genetics that show evolution has happened". Of these students, she was the most scientifically literate by virtue of her breadth of college course enrolment, but also as a matter of enthusiastic verbal fluency. Mary's deadpan dialogue plainly stated: "I think it's a valid scientific idea and has to do with many different aspects of science". Morgan describes a lived experience where the focus of my research generally (to better understand the source and shape of lived conflicts between students and others in learning environments) was mostly absent. When asked whether she had personally experienced science and faith conflicts at home or in school, she was mostly non-committal due to apparent lack of first-hand experience: "I've sort of seen some of that. I've taken two sociology classes since I've been here … they kind of play into that … but conflict over evolution? Not that I know of".

One slight outlier for this group of ecumenical students was Michael, who, while still both an advocate for science and faith in harmony societally, seemed to take a slightly more negative view of science: "Evolution?… It can really only be described as a theory. Some will argue that there's concrete proof; some will argue that there isn't".

### *Science in public*

While not a major point of discussion by these students, an appreciation of the health of science in the public sphere – and how it relates to their lives – punctuated some of our conversations. Naturally, given the mostly ecumenical nature of these students' outlook, the relationship between science and faith in public was an opportunity to speak positively. Typifying these kinds of interactions, Jackie discusses exploring trips to the local science museum:

> "My mom and dad took me to the local museum of natural history. I've gone there probably, like, five times. I think when my uncle went, he read us a lot of the things. I learned a lot of things from my uncle. He taught the application of evolution in like daily life … he's a chemical engineer."

Another distinct feature of conversations with such students was their marked appreciation of the social nature of group identification with specific issues. While this group were, notably, mostly apolitical, Mark explains a typical sentiment among these students regarding their views of the relation that faith and politics might have to someone's identity. Mark comments: "I do believe that political views are influenced by things like religion … if someone doesn't have a clear opinion on something, they tend to support the viewpoint of the party". Most of all, due to the pointed nature of many of their histories with classmates, school was by far the most common venue where appreciation of creationist sentiment was experienced.

### *Science and religion in school*

Unlike many creationists, for whom evolution in the curriculum of formal schooling poses personal problems (Long, 2011), students of a theologically moderate view had a distinctly different view of social conflict. Whereas the presence of evolution in their school's curriculum posed social tension for creationists, for these 'both/and' students, the presence of creationist complaint – and their sense of befuddlement at it – was distinctive. Students came from social milieus where, depending on the political valence of the debate within any one school, evolution may or may not have been taught meaningfully. As Berkman and Plutzer's (2010) longitudinal study of US biology classrooms showed, creationist complaint against evolution, while certainly more exacerbated in the US South, can be found almost anywhere if you look hard enough.

As Michael makes clear, for many students, conflict over science and faith in school is something not directly felt by these students, but experienced as a spectator: "When it came up in high school, you'd see people get really upset. I was just surprised by seeing people's reactions. People get really upset and heated about it". Elsa, another student, remembering her high school days, comments: "In high school, people would see that evolution caused some people to react and it would start debate … and other students would want to know why someone would have such a strong reaction to it". This was counter to the dominant theme running through their description of thinking about evolution. As most of these students described, evolution was interesting to them. Megan explains: "When we're on that topic in science, sometimes we'd talk about it with friends. We'd talk about how interesting it was how things turned out the way they did".

Almost every student in this pool of interviewees shared some similar story, if only mediated by the amplitude of social conflict. As some students recalled, the dynamics of conflict played out in slightly different ways, with varying consequences. Julie recalls the dynamics of her school:

> "In eighth grade, there was a growing debate in our school about the legitimacy of evolution … some of the more devout Catholic parents were trying to press the issue of creationism while the other side, of course, was more evolution-based learning. It grew to divide the class a little bit but I'm not sure how many of the students actually cared that much about it."

In the same social milieu of Roman Catholicism in the US, Morgan explains the more complex dynamics of evolution and faith within the same faith tradition and overlain upon schools. She explains how discussions over the relationship between science and faith surfaced for her:

> "There was tension starting early, which is really interesting because I went to a Catholic grade school. Now that I think of it, I think that may be the reason why that teacher isn't there anymore. Not that the school was forbidding the teaching of evolution … I just don't think it was their priority."

In a setting where Catholicism did not play as clear a role in students' social imagination, the same tension played out nonetheless:

> "Our teacher started with the fossil record, and the history of how scientists came to understand all of this and he said 'You don't have to believe all of this', and if I'm remembering correctly, he gave us the option to say at any time whether we were uncomfortable with the material at the time."

This opt-out mention by students, heard many times across this project, was seen as a customary accommodation to the curriculum. Further exacerbating this, some students describe the kind of evolution elimination found by Long (2011) and Berkman and Plutzer (2010). As Taylor states: "In high school, teachers weren't allowed to talk about it. In biology class, we skipped the chapter. I know we skipped the chapter". While a less common experience, a few students did detail a different valence to their evolution education in school: "I remembered studying natural selection in elementary school. I definitely remember doing that. That was my most impactful early science experience".

## Towards a different notion of the 'very religious'

The students described earlier, through their sense of religious self, or through their conceptualization of the religious 'Other' seen in the religious exclusivism they described, are, to my mind, not unusual. Few of them – in detail, really only two – described through the content of their interviews much more than a superficial understanding of their own or others' theology. Like the creationists I interviewed in the larger project (Long, 2011) within which this data were collected, both had precious little understanding of each other's faith. In my experience, only a minority of either identity had dexterous historical knowledge of each other. This seems to matter.

Tracing the history of Christianity from Judea through Western Europe, and eventually to its iteration in the US, is a well-known story.

Likewise, the rise of science in Western Europe, with its associated parallels in the Middle East and Eastern Asia, is also well known. Broad arcs of history, though, are largely the conceptual terrain of scholars – not lay citizens. The interaction between both science and faith, punctuating the histories of Newton, Descartes and Darwin (among many others), shows the relative historical political interdependence and, in the contemporary context, seeming independence of both as a social good. However, as the history of science and faith in the US has consistently shown, precious few Americans have such people ready at hand for interpreting the world. This kind of complaint is ameliorated by education, something the American system of formal schooling has historically come up short at when addressing knotty social interrelations.

As Canadian philosopher Charles Taylor (2007), perhaps more than anyone in recent memory, has shown, we have had our social imagination for the natural states of affairs regarding the appropriate roles of both science and faith in social life somewhat over-ensconced in our collective imagination by what is clearly not a natural state of affairs. History and culture – while both moving and changing – are contingent, not prescribed. Of the US's collective historical imagination, you might – depending upon with whom you speak – imagine Americans conjuring a story of divine providence and manifest destiny that animates a story of religio-nationalist development and expansion. This kind of animating narrative trope that many Americans carry with them does a certain kind of self-congratulatory work, but may be on the wane. In the early 21st century, Americans are increasingly two distinct things. On the one hand, religious observance overall is in decline (Pew Forum on Religion and Public Life and Pew Research Center, 2014). On the other hand, of those who are believers and practitioners, the majority of Protestants within them are increasingly represented by evangelicals and Pentecostals. Within the context of these social changes, what is the role of a fundamentalist subsection of the religious population within a social milieu with increasing immigration, religious diversity and the flourishing of cultural differences that represent deep, ontological differences? Are the students discussed earlier simply members of a shrinking pool of moderate Christian believers? Such social changes seem set to ineluctably amplify potential tensions over learning and about religious difference and the contingency of science as those who profess belief are increasingly aligned with a monological view.

We might, thinking of the religiously inclusive, entertain their relative openness across differing interpretive frameworks (both science and faith) as being a source for social amelioration. We can and should mobilize their discourse style as an explicit tool of social inclusion. Such inclusion, of course, is anathema to religious exclusivists as their proselytizing nature usually rules it out as reasonable action. If such religious amelioration is, in fact, a social good that we seek, then what we should be after is a way to move attitudes akin to the way we imagine religious fidelity. Currently, alongside the unearned virtue of the 'very religious', the forward looking inclusivist must leverage social action towards something more socially profitable. Of course, while often not attended to as diligently as could be, the social sciences' descriptive mission is often conflated with a normative one (see Kaden et al, Chapter 3, this volume) – we assume that the arguments we muster are assumed to be pointing naturally towards social improvement, even if the ends of that improvement are ideologically charged. As it currently stands, in the domain of US public education, there is almost no substantive dialogue at the interface of science and faith of the type I project being needed.

While the religious inclusivists' voices did not offer a depth of theological understanding across my work with them, the multiple voices with which they spoke showed a balance between, or at least humility towards, conceptions of truth that the religious exclusivists I worked with did not. Rather than a passive relation towards this state of affairs, I offer that a social and educative programme of religious and scientific literacy, with the explicit intention of complicating and ultimately tempering the ontological overreach of religiously exclusivist creationists, should be fostered within the US. The style of balanced reasoning employed by inclusivity is a virtue not to be lost, but is currently being marginalized and is in threat of disappearing due to demographic changes. If we see this view as a virtue worth retaining, we must take concrete steps. One such move would be to retract the seeming scalar-measured 'endpoint' status that fundamentalist versions of Christianity have acquired in the public's imagination of religious difference. With little more than history's happenstance behind it, the 19th-century's Holiness movements and other Christians of the Second Great Awakening – movements born out of England but flourishing in the US – and their historical denominational descendants have come to dominate what it means to be a Christian believer and practitioner in the early 21st century.

What I point at, in line with Smith and Holmwood (2013), is a conscious resurrection of using, as a social ameliorative tool, the public discourse style of the religiously moderate in service of advancing both science and faith in a pluralistic society. If current trends continue, though, the demographic shifts of dominant belief styles may make this a more fraught path in the coming years.

## Notes

1 Hill (2014) has suggested that some of this emphasis is an artefact of the emphasis of researcher questioning, which is often attracted to points of conflict, making it essentially a sampling bias problem.

2 Pseudonym.

3 Pope Pius XII's 1950 encyclical *Humanis generi* demarcated the work of God as inclusive of evolution whereas the social history of Roman Catholic understanding of evolution has clearly been marked by some anti-evolution attitudes among Catholics in the US.

## References

Allum, N., Sibley, E., Sturgis, P. and Stoneman, P. (2014) 'Religious beliefs, knowledge about science and attitudes towards medical genetics', *Public Understanding of Science*, 23(7): 833–49.

Berkman, M.B. and Plutzer, E. (2010) *Evolution, creationism, and the battle to control America's classrooms*, New York, NY: Cambridge University Press.

Bloom, H. (1993) *The American religion: The emergence of the post-Christian nation*, New York, NY: Simon and Schuster.

Borg, M.J. (2006) *Meeting Jesus again for the first time: The historical Jesus and the heart of contemporary faith*, Waterville, ME: Thorndike Press.

Butler, J., Mendieta, E. and Van Antwerpen, J. (2011) *The power of religion in the public sphere*, New York, NY: Columbia University Press.

Chan, E. and Ecklund, E.H. (2016) 'Narrating and navigating authorities: evangelical and mainline Protestant interpretations of the Bible and science', *Journal for the Scientific Study of Religion*, 55(1): 54–69.

Evans, M.S. (2016) *Seeking good debate: Religion, science, and conflict in American public life*, Oakland, CA: University of California Press.

Guhin, J. (2016) 'Why worry about evolution? Boundaries, practices, and moral salience in Sunni and evangelical high schools', *Sociological Theory*, 34(2): 151–74.

Hill, J. (2014) 'Rejecting evolution: the role of religion, education, and social networks', *Journal for the Scientific Study of Religion*, 53(3): 575–94.

Hunter, J.D. (1991) *Culture wars: The struggle to define America*, New York, NY: Basic Books.

Kahan, D.M. (2013) 'Ideology, motivated reasoning, and cognitive reflection: an experimental study', *Judgment and Decision Making*, 8(4): 407–24.

Long, D.E. (2011) *Evolution and religion in American education: An ethnography*, Dordrecht: Springer.

MacIntyre, A.C. (2010) *After virtue: A study in moral theory*, South Bend, IN: University of Notre Dame Press.

Marsden, G. (2006) *Fundamentalism and American culture*, New York, NY: Oxford University Press.

McGuire, M.B. (2008) *Religion: The social context*, Belmont, CA: Wadsworth Thomson Learning.

Numbers, R.L. (2007) *The creationists: From scientific creationism to intelligent design*, Cambridge: International Society for Science and Religion.

O'Brien, T.L. and Noy, S. (2015) 'Traditional, modern, and post-secular perspectives on science and religion in the United States', *American Sociological Review*, 80(1): 92–115.

Pew Forum on Religion and Public Life and Pew Research Center (2008) *US religious landscape survey 2008: Religious beliefs and practices; diverse and politically relevant*, Washington, DC: Pew Research Center.

Pew Forum on Religion and Public Life and Pew Research Center (2014) *US religious landscape study*, Washington, DC: Pew Research Center.

Pew Research Center (2010) *US religious knowledge survey*, Washington, DC: Pew Research Center.

Schnabel, L. and Bock, S. (2017) 'The persistent and exceptional intensity of American religion: a response to recent research', *Sociological Science*, 4(28): 686–700.

Smith, A.T.T. and Holmwood, J. (2013) 'Sociologies of moderation', *The Sociological Review*, 61(2): 6–17.

Swidler, A. (1986) 'Culture in action: symbols and strategies', *American Sociological Review*, 51(2): 273–86.

Taylor, C. (2007) *A secular age*, Cambridge, MA: Belknap Press.

Varenne, H. (1990) *Symbolizing America*, Lincoln, NE: University of Nebraska Press.

Wuthnow, R. (2008) *America and the challenges of religious diversity*, Princeton, NJ: Princeton University Press.

12

# Discourses on Science and Islam: A View from Britain

*Amy Unsworth*

## Introduction

In this chapter, I explore the discourses on Islam and science that are most prevalent in contemporary Britain. There is a relative lack of sociological literature on Islam/Muslims and science, and my intention in this chapter is to join some dots and highlight some gaps by providing the reader with relevant background, presenting some preliminary data and discussing problems and possibilities for future research.

I begin by considering the prevailing discourse among *non-Muslims*, but the majority of the chapter is dedicated to exploring discourses on science and Islam among Muslims in Britain themselves. Key to understanding the popular discourses on science and Islam is the importance of *da'wah* – Muslim missionary activity – and the South Asian reform movements influential in shaping Islam in Britain today. I therefore briefly outline the history of Muslims in Britain and then consider the views of reformist Muslim intellectuals and more recent popularizers who have been influential in the formation of 'Islam and science' discourses. I then present data gathered from focus group sessions conducted with Muslims of South Asian heritage who are actively engaged with Islamic institutions in Britain. Next, I use survey data to explore some of the themes raised in focus groups, and make suggestions for both qualitative and quantitative work on this topic in the future.

## Science and Islam: anti-religious discourses

'All the world's Muslims have fewer Nobel Prizes than Trinity College, Cambridge. They did great things in the Middle Ages, though' – so tweeted Richard Dawkins in August 2013.[1] He chose to tweet this on Eid al-Fitr, the most important religious holiday for Muslims worldwide, which marks the end of the month of Ramadan. The tweet prompted widespread criticism, with, for example, one commentator making the point that comparing a specialized academic institution to all the world's Indians, or all the world's Chinese, would also render the statement true and that this was 'agenda dressed up as fact' (Malik, 2013). Dawkins continued to defend his statement, tweeting several more times during the day in the same vein. In a follow-up post on his website, he claimed that this tweet was a reaction to the claims that 'the Qur'an anticipated modern science' and 'Muslim scholars kept the flame of Greek learning alight while Christendom wallowed in the Dark Ages'.

This was not the first time that Dawkins, a key figure in the New Atheist movement, famous for its pro-science and anti-religion stance, had made provocative statements about Islam. Earlier that year, he had tweeted that 'Islam is the greatest force for evil in the world today'. Sam Harris, another key figure in the New Atheist movement, dedicated an entire chapter in his book, *The end of faith* (Harris, 2005), to 'The problem with Islam'. Richard Dawkins's (2006) bestseller, *The God Delusion*, does not discuss Islam in detail, although he uses the slit in a burka as a metaphor to describe how science 'widen[s] the window', opening up 'so wide that the imprisoning black garment drops away almost completely, exposing our senses to airy and exhilarating freedom' (Dawkins, 2006: 406). The message here is clear: religion is a prison from which all people must be liberated through enlightenment, science is the tool of enlightenment and Islam is the archetype religion, the best representative of the darkness of religion.

This type of discourse, which asserts that the world's Muslims – all 1.8 billion of them – have not 'progressed' as they should have, is not new. Stephen LeDrew (2015: 75) has noted that Islam is an important element in the way that New Atheist writers construct their idea of an 'ideal Western civilization'. They appear to subscribe to ideas of cultural evolution that were prevalent in the 19th century, believing that all cultures follow the same trajectory and that there is some kind of problem with those cultures that have not developed in the same way as the West. Sam Harris (2005: 107) describes the Islamic world as a 'civilization with an arrested history'. The arguments made by contemporary New Atheists also echo the views of Ernest Renan, an influential 19th-century French scholar. Renan viewed European culture as the product of Semitic and Indo-Aryan cultures: while the Semites had given Europe religion, the Indo-Aryans had given it science and democracy. Since Islam remained untainted by other influences, it was considered the best example of Semitic monotheism to study, and epitomized many of the problems with Christianity, particularly Catholicism (Küçük, 2010). In 1883, Renan delivered an address entitled 'Islam and science' in Paris, in which he claimed that Islam had 'exterminated science down to the roots', concluding that 'what distinguishes the Muslim is a hatred of science, the conviction that research is useless, frivolous, impious: natural science, because it competes with God, and history, because it could revive the errors of the pre-Islamic period' (quoted in Amster, 2013: 64). Islam was being used as an exemplar to make an argument about the nature of the relationship between science and *all* monotheistic religion, although many scholars have subsequently read it as an attack on Islam alone (Küçük, 2010). The address was published in *Journal des Débats* in the same year and generated a transnational discussion among Muslim intellectuals from Russia, the Ottoman lands, Egypt and India (Aydin, 2007). Perhaps the most well-known refutation came from Jamal al-Din al-Afghani, to whom we shall return later in the chapter. Ironically, work such as Renan's 'became the source of much positive commentary on Arabic science and philosophy at the expense of medieval Christianity' (Küçük, 2010). In Draper's (1874) popular book, *History of the conflict between religion and science*, he draws on Renan and makes positive comments about Islam. For this reason, the work was translated into Turkish with an accompanying commentary to portray Islam – in contrast to Christianity – as the only religion compatible with science (Yalcinkaya, 2011).

The dominant discourse among anti-religious non-Muslim elites concerning science and Islam has certainly tended to be one of conflict, then. This type of discourse is also prevalent among more general publics: a recent study found that many participants, particularly those who were non-religious, viewed Islam as 'inimical to science', based on an assumption that Islam leaves no room for individual reasoning (Jones et al, 2019). However, the discourses among Muslims in Britain are rather different. In order to examine these, I first need to provide a little relevant background information.

## Muslims in Britain: a history and a typology

After the Second World War, large numbers of Muslims from the Indian subcontinent migrated to Britain. At first, mainly single men came, often via a 'chain migration' process, whereby a settler already employed in Britain would sponsor a male relative or friend to join him. This meant that many of the settlers came from the same few regions, particularly from the Sylhet region of Bangladesh and from rural villages in the Mirpur and Attock (formerly Campbellpur) districts of Pakistan. Most of these men expected to stay only temporarily, but from the late 1960s, women and children came to join their husbands and fathers, fearing that Britain's increasingly strict entry conditions would extinguish any possibility of family reunification in the future. With entire Muslim families now settling in Britain, it became far more important to establish facilities and services for Muslim communities. Less money was sent 'back home' and more was invested in establishing mosques and schools in Britain that would facilitate religious and cultural transmission to future generations (Gilliat-Ray, 2010).

Later still, Muslims from other parts of the world began to migrate to Britain, for example, from Turkey, Iran, several Arab states and Somalia. However, according to 2011 census data, the majority of Muslims in Britain today are still of South Asian heritage: 15% are Bangladeshi, 38% Pakistani and 7% Indian.[2]

Tariq Ramadan (2003) has suggested a categorization of Muslims according to six major tendencies, which I now briefly outline. I explain *Salafi reformism* in greater detail than the others because, as we will see, this trend has been influential in shaping Islamic discourses around Islam and science, as well as the perspectives of some younger Muslims in Britain.

Ramadan describes *scholastic traditionalists* as those who adhere to one of the main schools of jurisprudence: Hanafi, Maliki, Shaf'i or Hanbali (all Sunni), or Zaydi or Ja'fari (Shi'a). This means that interpretations of the Qur'an and Sunnah are only accepted from recognized scholars of the schools.

*Salafis*, in contrast, generally reject the traditional schools of jurisprudence and instead claim to follow the *Salaf*, 'the pious predecessors', that is, the first three generations of Muslims. Ramadan divides these, in turn, into three groups. The first is *Salafi literalists*, who insist on reference to the Qur'an and Sunnah and forbid any interpretive reading, which they condemn as *bidah* (innovation). They tend to function as isolationist communities and influential scholars are based in Saudi Arabia, Jordan, Egypt or Syria. Although the terms Wahhabi and Salafi are often used interchangeably, it is probably the Salafi literalist tendency that corresponds most accurately with Wahhabism. *Salafi reformists*, by contrast, are those inspired by Islamic thinkers of the late 19th and early 20th century, such as Jamāl al-Dīn al-Afghani, Muhammad 'Abduh, Rashid Rida, Muhammad Iqbal, Hassan al-Banna, Abul A'la Maududi and Sayyid Qutb (see next section). Although these thinkers differed in significant ways, they shared a desire to employ independent reasoning (*itjihad*) in interpreting the Qur'an and Sunnah to solve new societal challenges. They were therefore scripturalist but not literalist (Meijer, 2014). Two populist political movements that emerged from this reformist trend were the Muslim Brotherhood, founded by al-Banna in Egypt in 1928, and Jamaat-e-Islami, founded by Maududi (or Mawdudi) in India in 1941.

Salafi reformism, in Sadek Hamid's (2016: 18) words, 'developed as an intellectual response to Western ideological currents' and migrants brought this tradition with them when they settled in Britain in the 1960s. These ideas have been particularly influential in the subsequent creation of Islamic activist networks in the West and in efforts to re-Islamize Muslim young people. A distinctive feature of the religious institutions created by Salafi reformist activists in Britain is *da'wah*, which means 'invitation'. The invitation to Islam may be given to Muslims – to remind them of their religious obligations – or to non-Muslims (Hamid, 2016: 16). A narrative that promotes the scientific rationality of Islam is important to contemporary *da'wah* efforts, supported by a reading of the history of science that particularly emphasizes Muslim contributions, and a method of interpreting the Qur'an such that certain verses appear to pre-empt scientific discoveries. This type of interpretation is known as *I'jaz ilmi*, meaning 'scientific miraculousness'.

The final *Salafi* category in Ramadan's typology is *political literalist Salafism*, which describes those Muslims who desire to bring about radical social and political change inspired by a literalist reading of the texts. These latter two Salafi categories – *reformist* and *political literalist* – are commonly referred to as 'Islamist'.

The remaining categories in Ramadan's typology are *Sufis*, who emphasize the inner spiritual dimension of Islam and are typically members of a *tariqa*, or order, and *liberal or rationalist reformists* inspired by liberal and Enlightenment ideals. This latter group includes perspectives as diverse as the Pakistani scholar Fazlur Rahman and the UK-based 'counter-extremism' organization the Quilliam Foundation.

My study does not encompass these trends and so I will not describe them in further detail here.

## Salafi reformists and science

The Salafi reformists of the late 19th and early 20th century (mentioned earlier) were, in some ways, a disparate group, but they all promoted modernist agendas. These influential intellectuals tended to view Western science and technology favourably, in the sense that they felt Muslims needed to improve in these areas so that they might resist European colonialism. They attempted to harmonize the Qur'an with science in an effort to show that Islam was inherently modern. Particularly influential was Jamal al-Din al-Afghani, who called for Islamic unity to resist colonial rule and was highly critical of the Indian rationalist reformist Syed Ahmad Khan for his pro-Western stance. Al-Afghani also refuted Renan's views on Islam and science (see previous section). He conceded that Muslim countries were behind Western society at the time; however, he believed that Islam was a religion of reason and that the explanation for Muslims' current weakness was because they had neglected the rational faith of *al-salaf*, the pious predecessors. Therefore, returning to the original *Salafi* Islam was to be the remedy (Massad, 2015: 39–42).

Al-Afghani's follower, Muhammad 'Abduh, did more to address this issue and popularize the idea that modern European science is compatible with the Qur'an. He shared a revivalist ethos and also argued for naturalistic, non-miraculous understandings of events in the Qur'an (Sedgwick, 2014). Tantawi Jawhari was one of 'Abduh's main followers and is considered by some scholars to be the founder of the scientific interpretation of the Qur'an (Daneshgar, 2017: 58). He exhorted Muslims to learn science so as to advance the status of the Muslim world, and promoted the idea that discoveries made about the universe and nature can already be found in Islamic texts. He claimed that the Qur'an is science-based because it contains 750 scientific statements compared to just 150 statements on Islamic legal rulings (Abdul-Raof, 2012: 61). He concluded that Muslims therefore needed to reorder their priorities accordingly and place a greater emphasis on science. Jawhari also believed that Darwin's theory of evolution had previously been explained by medieval Islamic thinkers (Daneshgar, 2017).

Maududi (1903–79) was a highly influential figure who founded the organization and political party Jamaat-e-Islami in Pakistan. Maududi's vision was that Islam could provide a total system or social order, and he therefore wanted a truly Islamic state to be established. Born in India, he read Muhammad Iqbal, Ahmad Khan and other Muslim modernists, as well as Western intellectual thinkers (Jackson, 2010). Like the Wahhabis, he rejected the mystical Sufi elements in popular Islam. He was highly critical of those who dedicated too much attention to *fiqh* (Islamic jurisprudence) but was also vehemently opposed to the new Westernized elite of Muslim India – Sir Sayyid Ahmad Khan and those educated in the college he had founded. In an article published in 1936 ('al-Islam din al-ilm wal-aql' ['The religion of science and reason']), Maududi presented Islam as a religion that is based on science, and that can only truly be followed through science and reason. In contrast to Jawhari, however, he refuted Darwinian evolution, not by making arguments from scripture, but by attempting to make a scientific argument, claiming that because no 'Ape-man' remains had been found, Darwin's theory was nothing but 'inference and hypothesis' (Lerman, 1981). Maududi objected to the separation of spiritual education from worldly education, believing instead that all education should be Islamic:

> You teach them science which is devoid of reason and slave of the senses. You teach them economics, law and sociology which, in spirit and in substance, differ from the teachings of Islam. And you still expect them to have an Islamic point of view? (Abul Ala Maududi, Taalimat, quoted in Hoodbhoy, 2007: 53)

The belief among Salafi reformers that answers to modern problems can be found in the Qur'an arguably laid some of the groundwork for the development of *I'jaz ilmi*. This distinctive approach to interpreting the Qur'an was further developed by the French medical doctor Maurice Bucaille (1920–88), whose publications can be found in many Islamic bookshops and on *da'wah* stalls in the West. Bucaille was head of the surgical section at the University of Paris and was the family doctor of the Saudi king, Faisal bin Abdulaziz Al Saud (who died in 1975). His first book, *The Bible, the Qur'an and science*, was published in 1976 and translated into English in 1978. In it, he argues that the Bible contains numerous mistakes whereas the Qur'an is a genuinely infallible religious text. Bucaille referred to his own work in Egyptology as evidence, but he also gave other *I'jaz ilmi*-type examples, although it is unclear exactly what his influences were. His works were well received by many Muslims and inspired other publications in a similar vein; a *da'wah* film (*The book of signs*) was also produced to further propagate his claims. In a later book, *What is the origin of man?*, Bucaille dismissed as 'untenable' the idea that 'man was descended from the apes', suggesting that the interpretation of Qur'anic verses previously thought to be solely concerned with embryonic development should perhaps be extended to include 'the transformations of human morphology which took place over the ages' (Bucaille, 2005: 111).

Over the 20th century, the Saudi state played an important role in disseminating religious literature across Europe, South Asia and elsewhere, and these efforts also amplified Bucaille's ideas and *I'jaz ilmi* examples. In 1984, the Commission for Scientific Miracles of Qur'an and Sunnah was founded in Saudi Arabia by the World Muslim League. Among its objectives were to 'provide Muslim missionaries and mass media with Dawah' and 'to publicize the accepted researches … and to translate those papers into languages of the Muslim world and other living languages' (Iqbal, 2007: 163). The Commission's well-funded efforts have contributed significantly to the scientific-hermeneutic approach to the Qur'an and the popularization of *I'jaz ilmi da'wah* materials (Iqbal, 2007).

The works of Bucaille and other examples of *I'jaz ilmi* are referenced by the popular Indian Salafi preacher Zakir Naik, who founded the Islamic Research Foundation based in Mumbai and reaches huge audiences globally through his YouTube videos and his television channel, Peace TV.[3] Naik has delivered a lecture titled 'Is the Qur'an God's word?' numerous times, in which he insists that the divine authorship of the Qur'an can be proved because it anticipates modern scientific discoveries. He has also published the material in book form under the title *The Quran and modern science: Compatible or incompatible?* (Naik, 2008). In 2015, Zakir Naik received the prestigious Saudi King Faisal International Prize.

Zakir Naik's interest in Islam and science is in service to his primary aim of practising *da'wah* to strengthen the faith of Muslims. He has styled himself as an expert in 'comparative religion' and takes a strident approach to contesting other religions, particularly Hinduism and Christianity. Naik attempts to discredit other religions using scientific-sounding argumentation, while clearly placing science on the side of Islam. Coming from Mumbai, his style was forged in a postcolonial context where Muslims are in a minority and face a rising tide of Hindu nationalism (Kuiper, 2017). This means that his approach may particularly resonate with Muslims in other parts of the world who similarly find themselves as religious minorities facing significant pressures.

### *Harun Yahya*

In a discussion of discourses on Islam and science in Britain, it is necessary to briefly mention Harun Yahya's organization, which has produced a large amount of freely available English-language material dedicated to refuting Darwinian evolution and to promoting the idea that the Qur'an contains scientific miracles, claiming that:

> The Qur'an, which was revealed at a time when people knew very little about astronomy, physics, or biology, contains key facts on a variety of subjects such as the creation of the universe, the creation of the human being, the structure of the atmosphere, and the delicate balances that make life on earth possible. (Yahya, 2001: 11)

However, it is important to appreciate that Harun Yahya, based in Turkey, comes from a very different political context to that of the reformist *Salafis* previously discussed (for a detailed study of Harun Yahya, see Solberg, 2013).

## The study

The analysis presented in this chapter draws on preliminary data from three focus group sessions conducted with a total of 15 Muslims of South Asian heritage living in Britain (see Table 12.1 for further details of each focus group). Each focus group was intentionally fairly homogeneous in composition in order to allow me, the researcher and an outsider, better access to discourses commonly employed by participants (Hughes and DuMont, 1993).

One focus group was conducted with female family members in Bristol, a city in the South-West of England. The participants' parents had migrated from Pakistan to Britain, and these young women were now in their 20s; one was studying nursing and the others were working and married with children. They had been influenced to a great extent by their older sister, who had begun to take Islam much more seriously while studying at university. She had then introduced her sisters to Islamic scholars, books, blogs, study circles and events. Shortly after 11 September 2001 ('9/11') she had decided to start wearing hijab, to the consternation of her father, who was worried about her safety. Her younger sisters had all followed her example and chosen to wear hijab themselves. They followed the Hanafi School of jurisprudence and clearly distanced themselves from Salafi (literalism)/Wahhabism.

**Table 12.1: Focus group participants**

| Focus group code | Location | | Interviewee code | M/F | Age | Country of birth | Age arrived in UK | Ethnicity |
|---|---|---|---|---|---|---|---|---|
| FG01 | Bristol | Scholastic traditional – Hanafi | P07 | F | (20s[a]) | UK | n/a | Pakistani |
| | | | P08 | F | 24 | UK | n/a | Pakistani |
| | | | P09 | F | 20 | UK | n/a | Pakistani |
| | | | P10 | F | 28 | UK | n/a | Pakistani |
| FG02 | London | Salafi reformism | P14 | M | (60s[a]) | Bangladesh | – | Bangladeshi |
| | | | P15 | M | 23 | UK | n/a | Bangladeshi |
| | | | P16 | M | (20s[a]) | Bangladesh | – | Bangladeshi |
| | | | P17 | M | 18 | Bangladesh | – | Bangladeshi |
| | | | P18 | M | 27 | Bangladesh | – | Bangladeshi |
| | | | P19 | M | 41 | Sri Lanka | – | Sri Lankan |
| FG03 | London | Salafi reformism | P20 | F | 31 | Bangladesh | 1 | Bangladeshi |
| | | | P21 | F | 34 | Bangladesh | 9 | Bangladeshi |
| | | | P22 | F | 40 | Bangladesh | 8 | Bangladeshi |
| | | | P23 | M | 29 | Bangladesh | 28 | Bangladeshi |
| | | | P24 | M | (30s[a]) | UK | n/a | Bangladeshi |

*Notes*: [a] Age not given; age in brackets is an estimate; – = information not given; n/a = not applicable.

The other two focus groups were held at a large and influential mosque in London with significant historical links with Jamaat-e-Islami. One group was male-only and the other was mixed; both were mainly a mixture of first- and second-generation Bangladeshis. One particularly vocal member of the second group had studied a programme run by a pan-European Muslim organization, which was founded by former members of Jamaat-e-Islami. Recruitment to the first group was via a personal friend and recruitment to the other two groups was via a contact at the mosque.

Further attempts were made to conduct a focus group at another small neighbourhood mosque, but despite having a very helpful contact, unfortunately, sufficient people could not be persuaded to participate in the research. This means that Islamist/Salafi reformist views are particularly prevalent in the focus group data, and should not be presumed as representative of the entire Muslim population of Britain. This point is perhaps better made with reference to national survey data, which I shall discuss later in the chapter.

The focus group discussions were recorded and transcribed, and subsequently analysed using an inductive approach. I read the transcripts repeatedly to identify major themes and then coded the transcripts according to these themes. In the following analysis, I first explore aspects of Muslim identity formation within the British context. I then identify the discourses on Islam and science that emerged within the focus group discussions, explaining how this may relate to identity formation. Finally, I explore some of the information sources that resource active Muslims in Britain.

## *Being Muslim in Britain: "We're having to explain ourselves"*

In Focus Groups 2 and 3, which were both held at a Salafi reformist mosque, participants discussed how the experience of living as religious minorities and always being questioned about Islamic practices had resulted in what they perceived as a better-informed and strengthened faith:

> P19-FG02: "So, when you live with bunch of people, they just practising traditional way, you also try to do. So we come to countries like here. So everything has a question mark, why, how, when? Or, you question, then you come, come."

> P16-FG02: "It's from early age. You're questioned, isn't it. Because, 'Why do you have to eat halal?', for example. So, in order to answer that, you have to know why."
>
> P15-FG02: "as you become more independent and you start doing research and the reasons behind how this comes, you start looking at texts, like Qur'anic texts and Hadith texts and you see the evidence as to why you do this."

In Focus Group 3, one participant had recently emigrated from Bangladesh and described how it was only now that he was in the UK that he was making serious efforts to learn about Islam. Another participant in Focus Group 2 recounted a very similar experience when he first came to the UK from Bangladesh. Other participants, who had been in the UK from birth or from an early age, were of the impression that there were actually more resources available in the UK to learn about Islam than in Bangladesh. These views point to one possible reason for the appeal of Salafi reformism among British-born Muslims: compared to traditional scholastic forms of Islam, Salafi reformism seems to far better address this need to learn about Islam for oneself, driven, in part, by being questioned by non-Muslims.

While it was acknowledged within the group that it must have been hard for their parents' generation when they first arrived in the UK, it was also felt that things were perhaps even more difficult for their generation due to accusations of religious extremism:

> P24-FG03: "but, to an extent, they [our parents] weren't challenged about their faith. We're having to stand up for our faith and, er, you know, explain ourselves, you know, why we do this, why we do that, why we go to the mosque. It's like our religion is now under scrutiny and we're having to answer for it."
>
> P22-FG03: "I think we're caught in a position where if you're practising, you're questioned whether it's extreme and you're not integrating."

Two participants in this focus group clearly identified with a modern 'deculturated' (and globalized) form of Islam (Roy, 2006), making slightly disparaging remarks about their parents who "were taught in a different way" and identifying that "there were certain cultural practices integrated into religion which [are] non-founded". They stated that the "modern Muslim people" (ie not their parents) were "trying to become Islamic in the sense that it's right to be Islamic". However, other participants attributed much of their learning to their parents.

At one point during the conversation with Focus Group 1, one participant talked about the Big Bang theory, and this led to a conversation about evolutionary theory. There seemed to be consensus that they did not believe in evolution because they had no need for it. I then asked them about their experiences of learning about evolution at school. It was evident that the apparent contradiction with their religious beliefs had not been an important or salient issue for them at the time, with one saying, "to be honest, I just didn't put two and two together" (regarding the low salience of evolution as a topic among Muslims, see also Guhin, 2016). They continued:

> P08-FG01: "We didn't really take it on, it was just like, it's in the book, we read it, we got an exam on it [laughs].... We've been part of, like, two societies, you know, all our lives. It's not like we've lived in Pakistan and then we came here and we have to accept a whole new – so we, we've grown up with the theory that running alongside our religion and our family life, there is another culture, there's other religions out there, and to be understanding and accepting of them, you know. It's good, in fact, it's beneficial to know about other religions and what other people believe and what the theories are."
>
> P09-FG01: "And yeah, what's out there."
>
> P08-FG01: "It strengthens your own religion and your own faith."

Here, the participants demonstrate a relativistic attitude based on their experiences of growing up as British, Pakistani and Muslim. It is clear that they view evolutionary theory as a product of Western culture, and while they believe it is their duty to learn about and respect it as an alternative belief system, they do not need to accept it themselves – presumably mirroring their experience of other people not holding the religious beliefs that they themselves do. It is interesting to note that this encounter with difference ultimately serves to strengthen their own religious identity and beliefs. This example demonstrates a rejection of one aspect of Western culture, in contrast to the earlier examples in this section, which clearly demonstrate ways in which 1.5/2nd-generation participants rejected aspects of their parents' cultures.

### *Discourses on science*

During the focus group sessions, I asked participants to look at a list of several different career sectors, which included 'science', and to then discuss which career sectors might be particularly positive or negative work environments for a practising member of their faith. This exercise was designed to elicit initial thoughts about science without asking leading questions, and without introducing any particular framing of science and Islam, for example, one of conflict or of harmony.

The ensuing discussion among participants in Focus Group 1 revealed a perception that science is highly valued by Muslims and the role that parental aspirations may play in a young person's subject choices. The participants also demonstrated familiarity with both the history of Islamic science and the concept of *I'jaz*:

> P09-FG01: "'Science' is a big one for Muslims; they'd like to go into that a lot."
> INTERVIEWER: "And why is that?"
> P09-FG01: "I think because …"
> P08-FG01: "Because your parents tell you to! [Laughter.]"
> P09-FG01: "No, but I think like science in Islam as well, when you go way, way, way back, like, when you think of, like, drugs and things like surgical instruments and medicine and pharmaceuticals, Muslims, Arabs invented a lot of that."
> P10-FG01: "There's a lot of science in the Qur'an as well."

When asked to elaborate on the idea of science in the Qur'an, the participants gave examples of the sun and the moon,[4] and of the development of the human embryo in the womb.[5] During a discussion in Focus Group 3, one participant described science as "sometimes almost an extension of the Qur'an" and the group proceeded to give several examples of science in the Qur'an; while one person insisted that the Big Bang was described in the Qur'an, another disagreed. The development of the embryo was mentioned again, as was the earth's iron core.[6]

The discussion about science as a career sector in Focus Group 2 also had some similarities. The first response was a comment on how much science is contained within the Qur'an, ending with a generalizing statement that 'we' Muslims love to learn about science and again emphasizing the importance of Islamic science in the history of Western science. However, this participant also finished with a rueful acknowledgement that there is not a great deal of scientific research currently being carried out within Muslim societies. The only UK-born participant in this group immediately brought up the notion of conflict with Darwin's theory of evolution, but another participant quickly dismissed evolution as theory, not science, making a strong assertion that "science is always fact, proved". Interestingly, he believed that many scientists had proved that evolution is not true:

> INTERVIEWER: "And what about, final one I think, what about science?"
>
> P19-FG02: "Science is a – one third of Qu-, in Qur'an you get, OK, the science, out of more than 6,600 verses, 1,000 talk about science. Yeah, so we love to learn about science, if you get history, most of the modern world science, English, England, started from, er, Islamic world. So, therefore, we love to do that but Muslim society is not doing it! [Laughter.]"
>
> P15-FG02: "I think only conflict in terms of science things would be that, um, my understanding is stuff like maybe evolution theory, Darwin."
>
> P19-FG02: "That is not science. That is a theory. Science is always fact, proved. Evolution is hundreds and hundreds of scientists disproved it; there is no theory, there is no science at all."

The conversation continued, with another participant highlighting the issue of prejudice against Muslims in different career sectors, believing that less prejudice exists among the more highly educated, including in science. This view suggests that the participant is probably unaware of, or unbothered by, the New Atheist-type discourse of science–religion conflict and of particular hostility towards Islam. This participant also suggested that the source of any conflict between being a Muslim and doing science would lie in the realm of ethics, giving working in the arms industry as an example.

There was some further discussion about the theory of evolution in this group. One participant voiced serious concerns about the social implications of evolutionary theory, seeing it as underpinning a cruel capitalist system with no concern for the 'losers' in society. He also saw evolutionary theory as deeply problematic because he believed that it supported the notion of a racial hierarchy, with "white man" positioned at the top:

P19-FG02: "There's another big fact behind that. Big fact. I'm telling. Behind this evolution theory that's been promoted again by the capitalism. Secular system.... Survival of the fittest. If you are fit for the society, you are the winner. If not, you are the loser, you be kicked out. If you see that capitalist, capital society, I'm not completely against capitalism, right, there is the fact, balanced views there. Behind evolution theory, there's discrimination. There is a discrimination."

INTERVIEWER: "Discrimination?"

P19-FG02: "Yeah. There's a discrimination. You see that. So, you see that ape has been developed and then it's from the ape the black people came, then you can [laughter] yeah! That's the ..."

P16-FG02: "I mean, you could look at it like they haven't progressed. I mean, that's what it means, isn't it."

P19-FG02: "Yeah, finally they come to the white man! You can see the discrimination all there."

Another view of science and Islam was discussed in Focus Group 3. Here, Islam was described as itself being scientific:

P22-FG03: "But I think Islam is quite scientific because what science does to a person is to make them inquisitive and to research and test it out and to reflect."

P21-FG03: "So, it could become anything. Research. Put it to the test."

P22-FG03: "So, I think Islam as a religion actually gears somebody to do that. So you don't blind follow, you question, you reflect, you find, you, you know, you try to find ... not that you'll find answers to everything, um, and that's humanly not possible to find answers to everything, but it's to search for it and that's what science is, searching for certain things, for answers."

P20-FG03: "To prove."

P21-FG03: "To prove. Or disprove."

Here, the Salafi reformist perspective seems clear in the sense of advocating not only the idea that Islam and science stand in harmony, but also the idea that Muslims should not 'blindly follow', but search for answers. The participants then continued the discussion by talking about the health benefits of some Islamic practices, giving as examples the importance of ablution for good hygiene and of the health benefits of the Muslim positions of prayer and recommended position for sleeping.

## *Resourcing modern Islam in Britain*

A common theme in all the focus groups was that several decades ago, there had been very few Islamic resources available in English, but that the situation was now greatly improved. Participants in Focus Group 3 laughed together as they remembered studying the same one book:

> P22-FG03: "So, er, um, there's one particular book *Islam beliefs and teachings*, er, it'd have, I think that was the only child-friendly book around."
> P20-FG03: "At that time, yeah [laughs]."
> P22-FG03: "In the 80s [laughs] *The Ghulam Sarwar*?"
> P20-FG03: "Oh, that. The red cover. I think everyone probably …"
> P22-FG03: "Yes, yes. And that was really, I think really helpful. It was in English … and there weren't many books around in English at that time. So. But parental guidance a lot, in my case, I think."

This book was published by the Muslim Educational Trust, one of several organizations founded in Britain by Jamaat-e-Islami. It makes no mention of scientific content in the Qur'an, although it does contain an exercise that asks students to 'Contrast the Islamic understanding of creation with the theories developed by physicists and evolutionists to explain our existence'.

Participants in Focus Group 1 discussed the growth of Islamic TV channels available in Britain on Sky, which were 'unheard of' when they were growing up. When asked whether they had ever seen anything about science and religion in the media, they began talking about scientology (which was in the news at that time), but were unable to think of any other examples. Later in the session, when I asked them specifically whether they had ever heard of Harun Yahya, they all said yes, through their sister's books and through speeches and documentaries that they had seen on the Islam Channel. However, they agreed that they had "not come across any of his stuff" for quite a long time and one raised a slight concern about Yahya, saying "I think he's a slightly different type of Muslim to us … I think he's the Wahhabi or a Salafi or something". In Focus Group 3, Harun Yahya's name came up briefly during a discussion about public figures who addressed the subject of evolution, with one participant saying that she had read books by him "about mosquito and injection". Another said that Yahya was a "revert" (a convert to Islam). This is not the case, but it is possible that she was thinking of Maurice Bucaille, who wrote the widely circulated book *The Bible, the Qur'an and science*. The conversation then turned to a discussion about Usama Hasan, an imam who "actually says he believes in evolution". Although the participants themselves did not accept evolution, they expressed dismay and exasperation that Hasan had received threats after publicly supporting evolutionary theory.

The family members who made up Focus Group 1 talked about how they used to attend a lot of different events to learn about Islam:

> P09-FG01: "Before my sister got married and left then, I used to go with her, like, every event she used to go to, to learn about Islam. Um, you know, I used to go with her, we used to go to little, like, groups of people … houses … where we just sort of, like, remember God and stuff and pray."
>
> P08-FG01: "You know, there was a time in our lives when we were growing up, I think we were sort of hungry for knowledge and hungry to know more. We used to seek out events to go to and places to go and listen and learn. And I think since we've been so preoccupied with married life and family life and work, it's taken a bit of a back seat and we haven't really been to any kind of conferences."

Interestingly, one of the sisters clearly felt that a primary purpose of these events was to work out her identity as a British Muslim. Now, she said she "felt the balance", and therefore saw less of a need to attend these kinds of events.

They described how they used to travel to large events in cities like London, Leicester or Birmingham, and how the Islamic Societies at Bristol's two universities "are always running events". Mention was made of the 'Global Peace and Unity' event held at the London Excel Centre. This huge event, attracting over 50,000 visitors, was organized by the Islam Channel and was held most years from 2005 to 2013. A wide range of speakers featured in the programmes, including British politicians. It is now difficult to obtain a great deal of information about the events. The 2007 event featured an exhibition by Harun

Yahya, and the 2009 event featured Zakir Naik, who, as mentioned previously, promotes the idea of scientific miracles in the Qur'an. However, it appears that Naik did not speak about this subject specifically at the Global Peace and Unity event. (In 2010, Naik was banned from entering the UK.)

Many of the focus group participants said that they would use Google if they wanted to find out more about a topic related to Islam, with one individual joking that it was referred to as "Sheikh Google". Others said that they listened to audio lectures online or that there were "so many" videos on YouTube that they could watch. Some expressed a degree of caution about using online resources, aware that they would be likely to come across a very wide range of different viewpoints. A couple of people suggested that they would cross-check with offline sources, perhaps with a local scholar or a book by a known author.

## Survey data

Having analysed qualitative data from the focus group sessions I conducted with Muslims, I now turn to look at the results from a survey of views of science that I developed and carried out among 815 Muslims in Britain in 2014 with YouGov, as part of a larger study of science and religion.[7] As most of the survey questions were also asked of people of other faiths or no faith, it was not possible to explore some aspects of science and religion specific to Muslims. However, here, I present some of the data as a preliminary exploration of some of the focus group themes within a much larger population. In the discussion, I suggest how such survey work could be refined and extended in the future.

### *Learning science*

Focus group participants tended to place a very high value on science, with scientific and medical professions being highly respected. Figure 12.1 shows the percentage of survey respondents who studied various subject areas for their highest university qualification, comparing the Muslim subsample with a nationally representative sample. It is evident that Muslims are very much more likely to study medicine and health-related disciplines, computer/information science, and business; they are also somewhat more likely to study biological and physical sciences. Muslims are far less likely to study humanities or arts subjects compared to the general population. While

**Figure 12.1: Percentage of people studying various subject areas for highest university qualification**

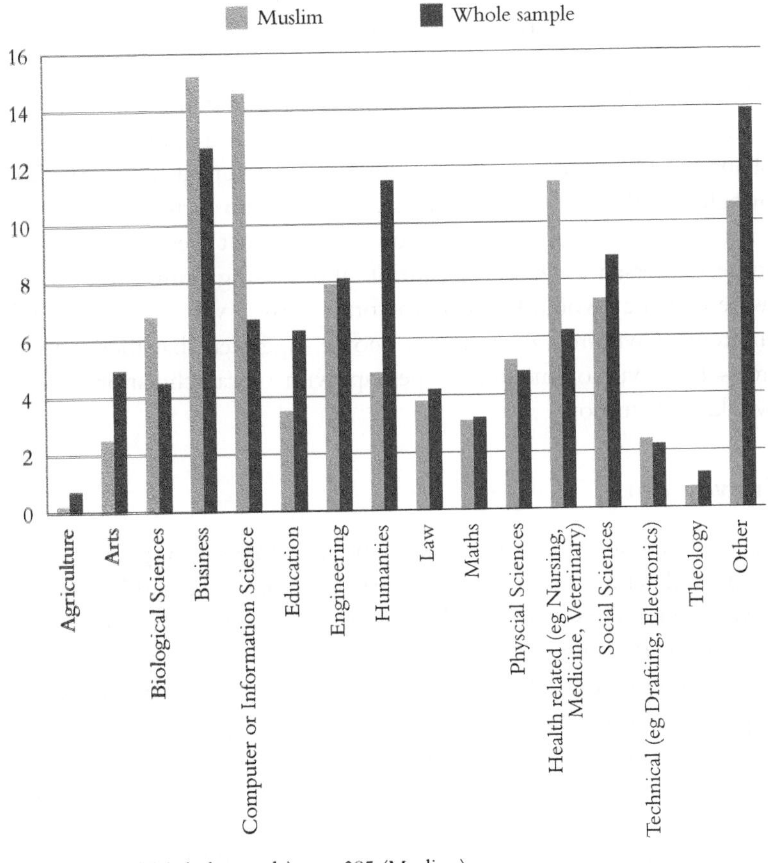


*Notes:* $n$ = 2,585 (whole sample); $n$ = 385 (Muslims).

it is true that, on average, Muslims in Britain show a preference for science-related disciplines, this may have far more to do with a desire for upward mobility and economic security than any explicitly religiously motivated reason.

## *Concerns about evolution*

The focus groups likely overstated the extent of opposition to evolution as while most of the focus group participants rejected evolutionary theory, I have previously shown (using the same survey data set of 815 Muslims) that 26% of Muslims reject plant and animal evolution and 44% reject human evolution. Furthermore, evolution does not appear to be a particularly salient issue among Muslims in Britain at large (Unsworth and Voas, 2018). It was notable that in one of the focus group discussions, distinctive objections to evolutionary theory emerged that would probably not be common among the most-studied American Christian anti-evolution groups, namely, concerns that evolutionary theory underpins a *laissez-faire* type of capitalism and the idea of white supremacy. Whereas only 5.1% of the general population agreed or strongly agreed with the survey statement 'the theory of evolution is racist', 18.4% of Muslims agreed or strongly agreed with it. A similar proportion of people from Black African or Caribbean backgrounds (mainly Christian) also agreed with the statement. Although it should be noted that greater proportions of these subpopulations did *not* agree that evolutionary theory is racist, it does raise the possibility that concerns among minority ethnic groups in Britain about evolution may have been unhelpfully reduced to the category of 'religion' and therefore delegitimized. The issue is, of course, complex given that a discourse linking evolution with racism is certainly promoted by some religious actors as a way of discouraging evolution acceptance.

## *Philosophies of science and support for science*

The 20th-century Muslim reformers who helped to shape some of the Muslim 'tendencies' in contemporary Britain tended to view science favourably, believing that Muslims needed better science and technology in order to improve their societies and compete with the West. This type of discourse was evident to some extent in the focus group sessions. The survey included a number of items exploring people's attitudes to science, as presented in Figure 12.2.

According to Figure 12.2, Muslims are slightly more likely to express an interest in science (Item 1). However, they have slightly less positivist views of science than are found among the general population: slightly fewer agree that 'theories in science can be proved to be definitely true' and slightly more agree that 'there are some things that science will never be able to explain'. There is less optimism among Muslims in Britain that 'science will help to make the world a better place in the future' and slightly less agreement that 'money spent on science is well worth spending'. Trust in scientists to generate reliable information is not particularly high among the general population or among Muslims: 57.8% and 56.5%, respectively, agree that they trust scientists. However, Muslims are rather more likely to agree with the statement that 'scientists too often let their own personal beliefs influence how they interpret their findings'.

**Figure 12.2: Percentage of Muslims or general population who agree or strongly agree with the statements about science**

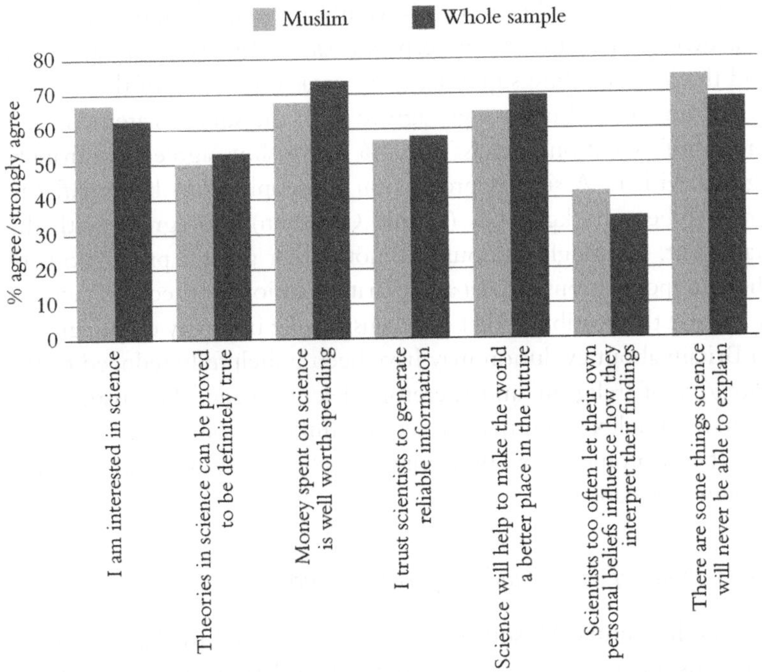


The overall picture is that Muslims in Britain do not differ very significantly from the general population when it comes to views of science. The differences between Muslims are greater; for example, Muslims who received their secondary school education (age 11–16) in the UK score lower on every item compared with those who were educated overseas (the majority of whom will be first-generation immigrants to Britain), thus revealing less positivist, less optimistic and slightly more critical views of science.

## Discussion

The activist efforts that began in the 1980s in Britain to re-Islamize Muslim young people drew a great deal of their inspiration from earlier Salafi Muslim reformists, who looked to science and technology and a renewed Islamic piety to bring about the modernization and decolonization of Muslim-majority lands. Focus group data suggest that in the present day, many Muslims in post-9/11 and 7/7 Britain, as members of a religious minority viewed with suspicion, may frequently find themselves in situations where they are required to 'explain themselves'. It is possible that the emergence of popular scientific atheism (or New Atheism) in the aftermath of 9/11, which depicts Islam as the archetype of a dangerous and irrational religious impulse, may have exacerbated this effect. Given this combination of factors, it is unsurprising that some Muslims in Britain draw on available *da'wah* discourses that centre Islam as rational, scientific and a source of resistance to oppression. Their significance is not primarily to persuade non-Muslims to convert to Islam, but to strengthen Islamic identity among Muslims who may be facing serious pressures. Focus group work would suggest that among Muslims actively engaged with religious media and institutions, these discourses on science and Islam are well known. However, initial survey work suggests that these views may not necessarily be widespread among the wider Muslim population in Britain. More research is required to establish this, in particular, qualitative work with a greater variety of Muslims, and in different regions of Britain. Sadek Hamid describes how it is not only Salafism that has been particularly influential among Muslims in Britain, but also varieties of neo-traditional Sufism influenced by figures such as US-based convert Hamza Yusuf. Are there recognizable discourses among Sufis with regards to science? Do they differ from those uncovered in this study?

It would also be extremely helpful to utilize a survey instrument specifically designed to interrogate views on science and Islam. Survey-based research of any minority group is always problematic given that it is extremely difficult to obtain a representative sample. Despite this, it would be advantageous to develop survey instruments that are better suited to surveying Muslim populations and testing the popularity of Islamic traditions and ideas such as those put forward in *I'jaz* narratives. In particular, it is important to recognize that the category 'Muslim' is very heterogeneous and it may be helpful to attempt to operationalize different Muslim tendencies.

This chapter also highlights the importance of being specific about which religion we are referring to in survey items. Very general statements such as 'science and religion are incompatible' may be straightforward for scientific atheists to respond to, but the dominant discourses examined here suggest that some Muslims may believe that it is only Islam that is compatible with science – perhaps some Christians believe the same about Christianity. It would also be interesting to discern whether different religions are viewed as differently in/compatible with science among non-religious populations. Relatedly, it would also be useful to survey the prevalence of New Atheist-type assumptions about Muslims among non-Muslim populations.

Many existing survey items related to evolutionary theory were developed and tested in the US and do not tend to include questions about the perceived social consequences of evolutionary theory, such as scientific racism and a capitalist economic system. However, it is important to recognize that opposition to evolution has different histories in different places, and our survey instruments need to better reflect this.

## Notes

1 See: https://twitter.com/richarddawkins/status/365473573768400896?lang=en

2 See the 2011 census for England and Wales, available at: www.nomisweb.co.uk/census/2011

3 A study that aims to catalogue all Internet videos addressing science and Islam found that Naik had created the highest number of videos on the subject (see Gardner and Hameed, 2017).

4 Participants did not elaborate any further, but it is likely that they are referring here to the 'scientifically miraculous' claim that the Qur'an pre-empts astronomical findings of the 16th/17th century showing that the earth orbits the sun. The claim is based on a particular interpretation of Sura 36:40.

5 This is an oft-cited *I'jaz* example, claiming that Sura 22:5 describes embryonic development in the womb in far greater detail than was humanly known at the time the Qur'an was written. Renowned Canadian embryologist Keith Moore is often referenced, who was invited to join the embryology committee of King Abdulaziz University in Jeddah to assist researchers with commenting on Qur'anic and Sunna phrases concerned with human embryonic development.

6 This claimed scientific miracle is based on Sura 57:25, which states 'we sent down iron, in it great might', and was written long before it was known that iron is formed in stars (as are many other elements).

7 The Muslim subsample (and other religious subsamples carried out as part of the survey) was not a probability sample, but was weighted for age, gender and social class to the 2011 census data for Muslims. For a more detailed description of the entire survey and sampling methodology, see Unsworth and Voas (2018).

## References

Abdul-Raof, H. (2012) *Theological approaches to Qur'anic exegesis: A practical comparative-contrastive analysis*, Abingdon: Routledge.

Amster, E.J. (2013) *Medicine and the saints: Science, Islam, and the colonial encounter in Morocco, 1877–1956*, Austin, TX: University of Texas Press.

Aydin, C. (2007) *The politics of anti-Westernism in Asia: Visions of world order in pan-Islamic and pan-Asian thought*, New York, NY: Columbia University Press.

Bucaille, M. (2005) *What is the origin of man?*, Delhi: Islamic Book Service.

Daneshgar, M. (2017) *Tantawi Jawhari and the Qur'an: Tafsir and social concerns in the twentieth century*, New York, NY: Taylor and Francis.

Dawkins, R. (2006) *The God Delusion*, London: Transworld Paperbacks.

Draper, J.W. (1874) *History of the conflict between science and religion*, New York, NY: Appleton.

Gardner, V. and Hameed, S. (2017) 'Science and Islam videos: creating a methodology to find "all" unique internet videos', *CyberOrient*, 11(1). Available at: www.cyberorient.net/article.do?articleId=9831.

Gilliat-Ray, S. (2010) *Muslims in Britain: An introduction*, Cambridge: Cambridge University Press.

Guhin, J. (2016) 'Why worry about evolution? Boundaries, practices, and moral salience in Sunni and evangelical high schools', *Sociological Theory*, 34(2): 151–74.

Hamid, S. (2016) *Sufis, Salafis and Islamists: The contested ground of British Islamic activism*, London: I.B. Tauris.

Harris, S. (2005) *The end of faith: Religion, terror, and the future of reason*, New York, NY: Free Press.

Hoodbhoy, P.A. (2007) 'Science and the Islamic world – the quest for rapprochement', *Physics Today*, 60(8): 49.

Hughes, D. and DuMont, K. (1993) 'Using focus groups to facilitate culturally anchored research', *American Journal of Community Psychology*, 21(6): 775–806.

Iqbal, M. (2007) *Science and Islam*, Westport, CT: Greenwood Press.

Jackson, R. (2010) *Mawlana Mawdudi and political Islam: Authority and the Islamic state*, Abingdon: Taylor and Francis.

Jones, S., Catto, R., Kaden, T. and Elsdon-Baker, F. (2019) '"That's how Muslims are required to view the world": race, culture and belief in non-Muslims' descriptions of Islam and science', *The Sociological Review*, 67(1): 161–77.

Küçük, H.B. (2010) 'Islam, Christianity, and the conflict thesis', in T. Dixon, G. Cantor and S. Pumfrey (eds) *Science and religion: New historical perspectives*, Cambridge: Cambridge University Press, pp 111–30.

Kuiper, M.J. (2017) *Da'wa and other religions: Indian Muslims and the modern resurgence of global Islamic activism*, London: Routledge.

LeDrew, S. (2015) *The evolution of Atheism: The politics of a modern movement*, New York, NY: Oxford University Press.

Lerman, E. (1981) 'Mawdudi's concept of Islam', *Middle Eastern Studies*, 17(4): 492–509.

Malik, N. (2013) 'Richard Dawkins' tweets on Islam are as rational as the rants of an extremist Muslim cleric'. Available at: www.theguardian.com/commentisfree/2013/aug/08/richard-dawkins-tweets-islam-muslim-nobel (accessed 2 October 2018).

Massad, J.A. (2015) *Islam in liberalism*, Chicago, IL: University of Chicago Press.

Meijer, R. (2014) 'Introduction', in R. Meijers (eds) *Global Salafism: Islam's new religious movement*, New York, NY: Oxford University Press, pp 1–32.

Naik, Z. (2008) *The Quran and modern science: Compatible or incompatible?*, Delhi: Adam Publishers.

Ramadan, T. (2003) *Western Muslims and the future of Islam*, New York, NY: Oxford University Press.

Roy, O. (2006) *Globalized Islam: The search for a new ummah*, New York, NY: Columbia University Press.

Sedgwick, M. (2014) *Muhammad Abduh*, Oxford: Oneworld Publications.

Solberg, A.R. (2013) 'The Mahdi wears Armani: an analysis of the Harun Yahya enterprise', Department of Literature, History of Ideas and Religion, University of Gothenburg, Sweden.

Unsworth, A. and Voas, D. (2018) 'Attitudes to evolution among Christians, Muslims and the non-religious in Britain: differential effects of religious and educational factors', *Public Understanding of Science*, 27(1): 76–93.

Yahya, H. (2001) *Miracles of the Qur'an*, Toronto: Al-Attique Publishers.

Yalcinkaya, M.A. (2011) 'Science as an ally of religion: a Muslim appropriation of "the conflict thesis"', *The British Journal for the History of Science*, 44(2): 161–81.

# Conclusion: Future Directions in the Sociological Study of Science and Belief

*Stephen H. Jones, Rebecca Catto and Tom Kaden*

One thing that this volume has made clear is that religious and other forms of ultimate belief interact with science in myriad ways, many of which are hardly recognized amid a public debate fixated on the simplistic categories of 'acceptance' and 'rejection' or 'conflict' and 'compatibility'. For different religious and non-religious communities of belief, 'science' can function as a cause to be fought for, a means of bolstering the appeal and authority of a revivalist movement, a means of making subtle cultural or class distinctions, or an imagined adversary to be challenged or competed with.

Among this volume's central aims has been to broaden the horizons of the study of science and belief, expanding upon the recent, innovative, US-based sociological research and moving the focus to states such as India, Pakistan, Turkey, Sweden, Canada and the UK. The further one moves afield, the more complex the subject becomes: it is clear that there is a whole world of difference between, say, the Midwestern US educational settings discussed by David Long in Chapter 11 and the elite contexts in Renny Thomas's study of Indian scientists in Chapter 6. Recent US-focused research has stridently moved away from the conflict thesis in the last decade, undermining the idea that public conflicts between conservative Protestants and secularist or scientific organizations are driven by knowledge differences. However, focusing the empirical lens on a country such as India – where it is unremarkable or even expected for elite scientists to see their findings and justify their work in religious terms – reveals this thesis to be provincial as much as it is simplistic.

Another aim has been to bridge the divides between the sociology of science, the sociology of religion and the emerging sociology of non-religion. All of these fields share an interest in questions of knowledge, belief and how these inform particular normative positions, and we remain convinced that there is a great deal that can be learnt from building firmer connections. There is, though, only so much that can be done to shape research agendas in one collection. In this conclusion, therefore, we want to not only consider themes emerging across the chapters, but also discuss some of the volume's limitations and use the insights found across it to make proposals for future work.

## Widening horizons

This collection has helped internationalize the sociological study of science and belief, but we would be the first to acknowledge that the steps we have taken here are only initial. While the book moves beyond the distinctive case of US, the majority of the chapters are focused on Western, historically Christian, contexts. The collection has been unable to address questions about how ultimate beliefs and science intersect in Latin American, South-East Asian or Central African contexts, or in Buddhist, Shinto, pagan, Sikh and other religious or secular traditions. The same can also be said of this field of research as a whole, although there are a few international examples among the growing body of work on science and religion presented and referenced in the volume (Cimino, 2014; Falade and Bauer, 2018). Here, again, sociology lags behind the disciplines of history and philosophy, where the study of, for example, Buddhism and science has a longer history and is considerably more developed (Lopez, 2010; Godart, 2017).

This collection does, however, suggest questions and points of focus for future investigation. The diversity presented across the foregoing chapters suggests that differing contexts pose differing questions and challenges, but there are also commonalities. For example, in chapters by Renny Thomas (Chapter 6), Amy Unsworth (Chapter 12) and Susanne Kind (Chapter 9), as well as in our own discussion of creationist and intelligent design movements (Kaden et al, Chapter 3), we have seen a theme repeated: individuals, from an array of religious and non-religious traditions and organizations, endeavouring to justify their ultimate beliefs by making reference to scientific concepts or by making a claim about what 'science' reveals to be true. This 'scientification of religion', as Kocku von Stuckrad (2014) terms it, is a surprising phenomenon considering that 'science', in the contemporary sense, only emerged during the 19th century, with students of nature previously thinking of themselves as pursuing 'natural philosophy' or 'natural theology' (Harrison, 2006). Questions about the extent of the cultural and societal dominance of science – the 'scientification' of society, as well as religion – emerge. How widely does this phenomenon extend? Is it only found in the Western contexts or the former Western colonies that we have covered here, or is it the case that the authority conferred by 'science' has genuinely 'gone global'? What are the implications for science and religion if the latter is increasingly justified in 'scientific' terms?

The chapters by Amy Unsworth (Chapter 12) and Jessica Carlisle et al (Chapter 7) also build a compelling case for an in-depth examination of Muslim contexts specifically. In their chapters, Unsworth and Carlisle et al show how, against the backdrop of the 'war on terror', unevidenced claims about Muslims' supposed opposition to science have flourished (Jones et al, 2019), and how evolution polling has been clumsily applied as a test of the 'modernity' of Muslim societies. There is perhaps justification for studying Muslim-majority and -minority contexts simply in order to challenge these decontextualized, and often sensationalized, studies. There are also several indications that the focus on Muslim perceptions of evolution, while of legitimate interest, has masked other, potentially more significant, intersections between Islam and science. The notion that science 'proves' the Qur'an's claims and validates its miraculous status, for example, appears to have considerable support across Muslim-majority and -minority contexts (Bigliardi, 2017; Moran, 2018; see also Unsworth, Chapter 12, this volume). Yet, sociological analysis of this trend is extremely scarce. It is, then, perhaps this, more than evolution, that should be the focus of sociological scholarship on Muslims and science. Broader postcolonial, critical secular studies work has deconstructed the powerful associations between modernity, the West and science, and the consequences of these for Islam and Muslims (Asad, 2003; Mahmood, 2015; Massad, 2015). Such focused research could also help theorize this and similar phenomena.

## Science and non-religiosity

One area that this collection has dedicated significant space to is questioning and analysing the apparent association between science and *non*-religion. While the non-Christian contexts listed earlier have suffered neglect due to their being – for anglophone scholars, at least – too remote and unfamiliar, this association has been neglected just as badly, but for the opposite reason that it is too close to home.

While it is now established that religious traditions have not always existed in opposition to the empirical study of nature, what is less often considered is the also commonly claimed natural affinity between non-religion and scientific progress. Lois Lee's claim in Chapter 8 that associations between science and non-religion are social and cultural – or even an accident of history – rather than 'natural' or purely epistemic thus invites deeper consideration.

Even if several chapters in this volume have covered non-religion and science, there are still, however, sizeable gaps that we have not filled in. We have seen in Stephen LeDrew and Susanne Kind's chapters (Chapters 10 and 9, respectively) that more organized varieties of humanism and atheism have been active in trying to define the boundaries of what 'counts' as science. As Lee pointed out, though, the perceptions of science of the majority of non-religious individuals who are *not* active in organized movements have rarely been examined. Also, as we observed in our Introduction, while various studies have explored whether knowledge of science affects religious opposition to scientific theories, almost never is this question turned around: do non-religious people who identify with science have a greater knowledge of or interest in it? If they do not, what kind of (social and cultural) factors motivate their identification? In our own work (Jones and Kaden, forthcoming), we have tentatively investigated this identification and found little connection between the two, but the systematic quantitative testing of non-religious science identification remains practically non-existent.[1] Finally, especially in light of Kind's and LeDrew's demonstration of relativism as a *bête noire* for organized non-religion in different national contexts, there is considerable scope for non-religion studies to engage in analysis of the status and boundaries of science. Such analysis ought to take place in conversation with the sociology of science, where analysis of the 'boundary work' done to define science is a well-ploughed furrow – albeit not in relation to existential cultures, as Silke Gülker demonstrated in Chapter 5.

## The science–religion binary

The final area that we wish to consider here is the categorization of science and belief, and how the categorizations utilized in research frequently promote the idea of conflict between science and religion, even unintentionally. We have seen various cases of this across this collection, some egregious, such as when evolution rejection is taken as a measure of strong belief, or acceptance is taken as a marker of disbelief. However, even if one sidesteps these clear pitfalls, others are harder to avoid. In Chapter 5, Silke Gülker pointed out, rightly, that research and reflection on the relationship between science and religion tends to take the two concepts themselves largely at face value, despite the fact that defining what exactly 'science' is or what 'religion' is remains incredibly difficult and is, in the case of both terms, a matter of considerable scholarly debate. Avoiding these terms entirely in survey or interview research is, however, next to impossible – so it is hard to avoid becoming weighed down by firmly established cultural narratives. This is made clear by David Long in Chapter 11. Long rightly argues that there is no need for researchers to assume what he terms a 'sliding scale' from fundamentalist creationism, to theistic evolutionism, to atheistic evolutionism. Indeed, this approach is profoundly misleading since it appears to suggest that the Catholic opponents of evolution in Long's study are 'more religious' than the Pope. Yet, Long also showed that this way of arranging categories is difficult to escape: even those religious people who accept evolution, and oppose creationist discourses, still often credit their adversaries with being 'stronger' in faith.

While such problems are hard to escape, there are perhaps three ways in which researchers of science and belief can protect themselves and even help limit the risk of reifying pre-existing categories and stereotypes of science and religion in wider public life. Several of the chapters in this volume – including Hill's (Chapter 2), Reid's (Chapter 4), Carlisle et al's (Chapter 7) and our own (Kaden et al, Chapter 3) – made it clear that developing survey questions on science and belief and interpreting answers is not straightforward: changing the question, even subtly, can lead to a different story about science and belief being told. One shift, in particular, that both our and Hill's chapter called for in response to problems in past surveys was for research instruments to not ask directly about the categories 'science', 'religion' or even 'evolution', but to probe related issues, such as people's views on human consciousness,[2] or what people believe 'science' to involve practically or imply philosophically. This is not to say that survey instruments directly asking people if 'science' and 'religion' exist in conflict are not useful for highlighting common perceptions. Rather, it is to suggest that focusing less on whether people believe *in* subjects like evolution and more on what they believe *about* them, as recent research has done (Hill, 2014; Elsdon-Baker, 2015; Elsdon-Baker et al, 2017; McCain and Kampourakis, 2018), could helpfully disrupt pre-framed categories, allowing them to be seen in a new light. Unsworth's use of a discussion prompt on science career pathways in focus groups described in Chapter 12 offers a helpful qualitative example of how to interrogate science and belief issues more obliquely.

Alongside this, very few studies of science and belief consider the impact of public media on people's perceptions. Typically, people's views about science and religion, especially religious people's views, are thought to be shaped through immediate networks, especially the spheres of church and home. In Long's research on religious moderates, however, we can glimpse the possibility that people's 'horizon of understanding' may be influenced by how religion and science are presented in popular media.[3] Such is also suggested by other unusual findings covered in the preceding chapters, such as the finding by Pew (discussed in Chapter 7) that American Muslims seem to reject evolution as a higher rate than Muslims living in contexts where Islam is more influential. This is a finding echoed in some of Unsworth's Muslim focus group participants' accounts of feeling 'under scrutiny' as a minority in Britain and the consequent need to learn and understand more about their own faith and how it connects to other world views.

Finally, a theme that emerges across chapters in this volume is that religion should not be given privileged status in analysis of perceptions of science relative to other aspects of individual identity. Carlisle et al, for example, highlighted in Chapter 7 the need to approach Muslims' identities intersectionally: questions about science acceptance need to be considered in light of class, ethnic, gender, political and other differences, and not just in light of their religious commitments. This was clear, too, in Thomas's chapter, which highlighted how, in India, 'being scientific' remains bound up with religious, class and ethnic distinctions. Intersectional research into science and religion, which has only recently begun to appear even in the US (Noy and O'Brien, 2018), could not only help undermine stereotypes about science and belief, but also bring fresh insights into contemporary inequalities related to science and belief internationally. This means considering religion and belief seriously, but without seeing it as the only relevant factor.

## Towards an integrated treatment of science and belief

For all these areas to move forward, one thing that will be necessary, as we observed in the Introduction, is greater overlap between different areas of the sociological discipline. This is difficult, however, not just because of the usual practical challenges that those working across academic specialisms face, but because it means questioning the foundational assumptions of sub-disciplines. The aforementioned calls to pay greater attention to how both religious and non-religious people identify with, and make claims on, science, for example, are also calls for a shift in focus within the wider field of the public understanding of science – which, as Fern Elsdon-Baker and Will Mason-Wilkes in Chapter 1 and Jessica Carlisle et al in Chapter 7 make clear, has tended to focus more narrowly on trust in, knowledge of and access to scientific ideas and institutions. Similarly, engaging with this theme in the sociology of religion means shifting away from the idea of religion as a *sui generis* category and looking at how ultimate meaning emerges in a variety of knowledge traditions.

Such questioning is ultimately worthwhile because inaccurate, overgeneralized portrayals of conflict between science and religion can feed misunderstandings and negative stereotypes (Jones et al, 2019). In Chapter 5, Gülker quoted from Merton's classic study of science and Puritanism in 17th-century England: '[W]e are concerned with the complex modes of interaction between a religious ethic and science, not as these appear to apologists of the two camps, but as they occurred in the course of actual social development'. Organizations and individuals seeking to promote a specific agenda can and do continue to exploit emic categories and stereotypes today, and scholars need to respond accordingly. A more international, reflexive and integrated sociological approach eschewing pre-existing assumptions about science and religion (eg that publics' responses will fit on a sliding scale from conflict to harmony) is needed if the conflict thesis is finally to be put to bed and new agendas enabled to flourish.

## Notes

1. Related findings from the broader 'Science and Religion: Exploring the Spectrum' project are also in development. For more information about this project, see our Introduction and Chapter 3.
2. Such a focus on human consciousness is prominent in one new survey by Elsdon-Baker et al (2017).
3. Michael S. Evans (2016) is another rare example, analysing connections between public debate on science and religion and publics' perceptions of and desires for such debate in the US.

## References

Asad, T. (2003) *Formations of the secular: Christianity, Islam, modernity*, Stanford, CA: Stanford University Press.

Bigliardi, S. (2017) 'The "scientific miracle of the Qur'ān", pseudoscience, and conspiracism', *Zygon*, 52(1): 146–71.

Cimino, R. (2014) *Mystical science and practical religion: Muslim, Hindu, and Sikh discourse on science and technology*, Lanham, MD: Lexington Books.

Elsdon-Baker, F. (2015) 'Creating creationists: the influence of "issues framing" on our understanding of public perceptions of clash narratives between evolutionary science and belief', *Public Understanding of Science*, 24(4): 422–39.

Elsdon-Baker, F., Leicht, C., Mason-Wilkes, W., Preece, E. and Piggott, L. (2017) 'Summary report of preliminary findings for a survey of public perspectives on evolution and the relationship between evolutionary science and religion'. Available at: https://sciencereligionspectrum.org/wp-content/uploads/2017/09/SRESYouGov-survey-preliminary-findings-5.9.17.pdf (accessed 14 April 2018).

Evans, M.S. (2016) *Seeking good debate: Religion, science, and conflict in American public life*, Oakland, CA: University of California Press.

Falade, B.A. and Bauer, M.W. (2018) '"I have faith in science and in God": common sense, cognitive polyphasia and attitudes to science in Nigeria', *Public Understanding of Science*, 27(1): 29–46.

Godart, G.C. (2017) *Darwin, dharma, and the divine: Evolutionary theory and religion in modern Japan*, Honolulu: University of Hawaii Press.

Harrison, P. (2006) '"Science" and "religion": constructing the boundaries', *The Journal of Religion*, 86(1): 81–106.

Hill, J.P. (2014) *National study of religion and human origins*, Grand Rapids, MI: BioLogos Foundation.

Jones, S.H. and Kaden, T. (forthcoming) 'Science and religion as lived experience: narratives of evolution among British and Canadian publics and life scientists', in B. Lightman and F. Elsdon-Baker (eds) *Science and religion: Exploring the spectrum*, Pittsburgh, PA: University of Pittsburgh Press.

Jones, S.H., Catto, R., Kaden, T. and Elsdon-Baker, F. (2019) '"That's how Muslims are required to view the world": race, culture and belief in non-Muslims' descriptions of Islam and science', *The Sociological Review*, 67(1): 161–77.

Lopez, D.S. (2010) *Buddhism and science: A guide for the perplexed*, Chicago, IL: University of Chicago Press.

Mahmood, S. (2015) *Religious difference in a secular age: A minority report*, Princeton, NJ: Princeton University Press.

Massad, J.A. (2015) *Islam in liberalism*, Chicago, IL: University of Chicago Press.

McCain, K. and Kampourakis, K. (2018) 'Which question do polls about evolution and belief really ask, and why does it matter?', *Public Understanding of Science*, 27(1): 2–10.

Moran, G. (2018) 'British Muslim perceptions of evolution', PhD thesis, Liverpool Hope University, UK.

Noy, S. and O'Brien, T.L. (2018) 'An intersectional analysis of perspectives on science and religion in the United States', *The Sociological Quarterly*, 59(1): 40–61.

von Stuckrad, K. (2014) *The scientification of religion: An historical study of discursive change, 1800–2000*, Boston, MA: Walter de Gruyter Inc.

# Index

## Q



## R



## S

## T



## U

Printed and bound by CPI Group (UK) Ltd, Croydon, CR0 4YY

18/09/2026

14965016-0004